Modern Graph Theory Algorithms with Python

Harness the power of graph algorithms and real-world network applications using Python

Colleen M. Farrelly

Franck Kalala Mutombo

Modern Graph Theory Algorithms with Python

Group Product Manager: Ali Abidi

Publishing Product Manager: Yasir Khan

Book Project Manager: Hemangi Lotlikar

Senior Editor: Tazeen Shaikh

Technical Editor: Rahul Limbachiya

Copy Editor: Safis Editing

Proofreader: Tazeen Shaikh

Indexer: Subalakshmi Govindhan

Production Designer: Jyoti Kadam

DevRel Marketing Coordinator: Nivedita Singh

First published: June 2024

Production reference: 1230524

Published by Packt Publishing Ltd.
Grosvenor House
11 St Paul's Square
Birmingham
B3 1RB, UK.

ISBN 978-1-80512-789-5

www.packtpub.com

Many thanks and much gratitude to Peter Schnable, for his encouragement and guidance during my journey into science and mathematics; for long discussions about herpetology, conservation, and ethics; and for inspiring me to share my knowledge with others and to forge my own path in STEM for social good.

- Colleen Molloy Farrelly

To my beloved mother, Ngoy Justine, whose unwavering love and encouragement fueled my passion for knowledge. Throughout the writing of these pages, her memory has been a guiding light, inspiring me to pursue excellence and share the joy of discovery. Though she is no longer with us, her spirit lives on in the words and sentiments expressed within these chapters. With profound gratitude and love, I dedicate this work to the woman whose influence continues to shape my journey.

To Meda, Divine, and Abigael.

- Franck Kalala Mutombo

Foreword

As the CTO/CIO of life sciences as well as automotive, energy, and high-tech industry-focused companies, I have continuously been challenged with creating meaningful data insights from a great variety of unstructured and semi-structured data sources. A decade ago, my advanced analytics and machine learning journey experienced the biggest push forward when I benefited from one of the most impactful introductions in my career. Ever since I got to work with Colleen Farrelly on our ontology-focused data science requirements when building out a genomics diagnostics platform, my approach to AI has been tremendously elevated. Subsequently, I got to work on Colleen Farrelly's exceptional book, The Shape of Data, which intrigued me deeply with her unique delivery of how to combine geometry and machine learning-based algorithms and supporting data structures to create powerful topological representations of complex data problems.

In this equally captivating sequel, Modern Graph Theory Algorithms with Python, Colleen Farrelly and Franck Kalala Mutombo take you to the next level of unleashing the potential of network science by diving deep into specific graph theory approaches to solve a great variety of industry problems ranging from ecological to financial, spatial and temporal sales, and clinical data challenges. It is easy to see why readers and data science practitioners of any level will find each chapter profoundly valuable based on the supporting Python library and code examples along with the underlying mathematical explanations.

With each turn of a page, I found myself wanting to reinforce the presented learnings by applying well-defined examples to my own business requirements. Being a keen user of graph databases, I very much enjoyed the hands-on practical discussions on specific technologies such as in *Chapter 12* to illustrate the relative ease of Python integrations and built-in capabilities of open source solutions that can be leveraged in today's growing powerful set of available tools.

This comprehensive book provides just-enough and just-in-time fundamental concepts to enable data scientists and software engineers to greatly elevate their machine learning techniques with directly applicable well-structured scenarios. Each of the consistent problem-solution breakdowns includes key considerations for the required data wrangling, transformation, and modeling aspects.

Furthermore, this book provides a convincing lead into the relevance of new frontiers such as quantum network science algorithms, neural network architectures as graphs, hierarchical networks, and hypergraphs. With a concise and easy-to-follow thought process, this book provides you with the important context of how to reduce the large volumes of parameterization required for large language models and address the critical aspect of metadata management via hypergraph databases, for example.

The use of graph theory today is highly relevant to every industry and science domain. Whether the challenge is to provide predictive modeling or simulations or the optimization of business operations or clinical outcomes and many more requirements, this book is an indispensable guide to mastering the complexities of these critical real-world challenges. With all the key insights and GitHub repository examples at your fingertips, you will be transformed instantly into a subject matter expert. The Modern Graph Theory Algorithms with Python exploration is a must-read and thoroughly enjoyable book.

Michael Giske

Chairman of Inomo Technologies and Global CIO of B-ON

Contributors

About the authors

Colleen Molloy Farrelly is a chief mathematician, data scientist, and researcher who has expertise in applying math to the biological, medical, social, and physical sciences. She has also authored the book, *The Shape of Data*. She has mentored, coauthored papers, and worked with people across Latin America, Africa, Europe, and Asia.

She is based in Miami, Florida in the US and holds a master's in biostatistics from the University of Miami. She is passionate about educational initiatives in the developing world and speaks at conferences such as Women in Data Science, IEEE conferences, PyData, and Applied Machine Learning Days.

I want to thank the people who have supported me over the years, especially early on, including John and Nancy Farrelly, Peter Schnable, the Warmus family, Mr. and Mrs. De Jong, the Mayor families, Justin and Christy Moeller, Luke Robinson, and many professors and colleagues throughout my career.

To all my students and those who will come after me who motivate me to teach and share my knowledge, you can use math and science to change the world for the better.

Franck Kalala Mutombo is a professor of mathematics at Lubumbashi University and former academic director of AIMS-Senegal. He previously worked in a research position at the University of Strathclyde and AIMS-South Africa in a joint appointment with the University of Cape Town. He holds a PhD in mathematical sciences from the University of Strathclyde, Glasgow, Scotland. His current research considers the impact of network structure on long-range interactions applied to epidemics, diffusion, object clustering, differential geometry of manifolds, finite element methods for PDEs, and data science. Currently, he teaches at the University of Lubumbashi and across the AIMS Network.

I express gratitude to my supportive network throughout my journey. I'm thankful for friends and professors who've contributed to my career. I am indebted to the countless students across Africa and those who will succeed them. Their enthusiasm and curiosity serve as constant reminders of the profound impact that mathematics and science can have on shaping a better world. It is with gratitude that I embrace the opportunity to teach and share knowledge, fostering a community of learners committed to leveraging the transformative potential of these disciplines for the greater good.

About the reviewer

Casey Moffatt, with a master's in applied mathematics and a double bachelor's in pure mathematics and philosophy, specializes in graph theory, optimization, and computer science. He is proficient in Python and various essential software for graph data science, machine learning, and algorithm development. He is eager to push boundaries in mathematics and computer science. He would like to thank Packt Publishing and contributors for enabling projects like this and to the countless individuals behind open source technologies.

Table of Contents

Part 2: Spatial Data Applications

3

4

5

Part 3: Temporal Data Applications

6

Stock Market Data   101

7

Goods Prices/Sales Data   121

8

Dynamic Social Networks   137

Part 4: Advanced Applications

9

10

11

Mapping Language Families – an Ontological Approach   197

12

Graph Databases   211

13

Putting It All Together   229

14

Preface

Hello there! **Network science** combines the power of analytics with the deep theoretical tools of graph theory to solve difficult problems in data analytics. This empowers researchers and industry engineers/ data scientists to analyze data at scale and reframe intractable analytics problems to produce powerful insights into problems and predictions about system behaviors, including biological, physical, and social systems of interest.

There are many important applications of network science today, including these:

- Social network data

- Spatial data

- Time series data

- Spatiotemporal data

- More advanced data structures, such as ontologies or hypergraphs

This book gives a brief overview of social network applications and focuses on the cutting edge of network science applications to areas of data science, such as transportation logistics, conversation, public health, linguistics, and education. By the end of your journey, you'll be able to frame your own data problem within the framework of network science to derive insights and tackle difficult problems in your field.

We will provide the necessary mathematical background as we dive into practical examples and code related to our work in academia and industry over the past decades, including work on predicting Ebola outbreaks, forecasting food price volatility, modeling genetic and linguistic relationships, and mining social networks for insights into social tie formation. As the world faces food shortages, public health crises, economic inequality, supply chain breakdowns, and environmental crises, network science will play an important role in big data analytics for social good.

Who this book is for

This book is for you if you are working with data. To get the most out of the book, you should have some familiarity with Python, particularly the `pandas` and `numpy` packages. In addition, some familiarity with data analytics is assumed, though the network science tools and problems we tackle are built from scratch for readers without a background in those problems or methods.

Network science has a rich history in many scientific disciplines, including epidemiology, biomedical engineering, sociology, genetics, environmental science, particle physics, computer science, and economics. Its foundations in graph theory influence research in many areas of pure and applied mathematics as well. Anyone in the fields of science, technology, engineering, and mathematics can benefit from network science's toolset and approach to problem-solving.

What this book covers

Chapter 1, What Is a Network?, introduces the theoretical concept of a network and provides several examples of networks in real-world applications, including work with random graphs. We'll also get started with Python's igraph and NetworkX packages.

Chapter 2, Wrangling Data into Networks with NetworkX and igraph, builds on *Chapter 1* by providing three examples of real-world data that can be formulated as network data and showing how to convert data into network form in Python. We'll introduce problems involving spatial data, temporal data, and spatiotemporal data and explore how network science can solve these problems by converting the data into network form.

Chapter 3, Demographic Data, explores two real-world projects using demography data from the developing world to understand network structures and capacity for information/infectious disease spread. We'll consider the demographic characteristics and network properties of a friend group to see how both types of information can influence disease spread.

Chapter 4, Transportation Data, provides a real-world example of a transportation network and introduces tools related to minimum paths and network flow. We'll consider optimal routing and the shortest paths to destinations, including multistop pathways from one location to another.

Chapter 5, Ecological Data, shows a real-world example of an ecological network and introduces spectral graph theory tools, including spectral clustering and graph Laplacians.

Chapter 6, Stock Market Data, examines a real-world example of stock market data analysis with network tools, including edge-based centrality measures of volatility. We'll mine data for tipping points, heralding either a period of market growth or market crash.

Chapter 7, Goods Prices/Sales Data, provides two real-world examples of commerce data analysis over both space and time with tools previously covered in time series and spatial data applications. We'll examine sales and pricing trends across time and space to better understand consumer behavior and the impacts of pricing changes across time and space.

Chapter 8, Dynamic Social Networks, introduces a real-world example of social network datasets evolving over time and analyzes their vulnerability to spreading processes, such as epidemics and misinformation flow. We'll consider factors influencing ecological social networks' vulnerability to spread of disease.

Chapter 9, Machine Learning for Networks, presents a comprehensive description on network-based machine learning and deep learning, including examples with supervised, unsupervised, and semi-supervised learning to understand disease risks within social networks.

Chapter 10, Pathway Mining, introduces Bayesian networks and mining for causal pathways using an educational data example, where we'll see how course sequencing and performance influence student outcomes.

Chapter 11, Mapping Language Families – an Ontological Approach, covers ontologies and maps between ontologies using a linguistic data example from the Nilo-Sudanic language family and its lexicon variations.

Chapter 12, Graph Databases, introduces graph databases with Neo4j, including data from prior chapters and how to query Neo4j with graph tools introduced in prior chapters and Neo4j's query language. We'll see how graph databases and network science tools create synergy in data science, as well as efficient data storage solutions.

Chapter 13, Putting It All Together, ties together material from previous chapters into a final project, analyzing spatiotemporal network data and demographic data from Ituri and North Kivu provinces with generalized estimating equations to understand the evolution of the 2019 Ebola epidemic.

Chapter 14, New Frontiers, introduces quantum graph algorithms, graph theory for neural network optimization, hierarchical networks, and hypergraphs.

To get the most out of this book

We provide Python scripts and assume some knowledge of basic Python analytics packages (such as NumPy and scikit-learn) and Python syntax. We assume some knowledge of basic analytics tasks such as summary statistics and working with different types of data in Python with either `numpy` or `pandas`. Scripts are written for each chapter, with later scripts often depending on earlier scripts in the chapter to build knowledge. Other concepts in Python and in analytics will be introduced conceptually and then with Python code examples.

Software/hardware covered in the book	Operating system requirements
Python 3.12.3	Windows, macOS, or Linux

If you are using the digital version of this book, we advise you to type the code yourself or access the code from the book's GitHub repository (a link is available in the next section). Doing so will help you avoid any potential errors related to the copying and pasting of code.

You are encouraged to try out the code examples in this book on your real-world data science projects. If you want to delve deeper into graph algorithms and network science, we encourage you to look at the latest research papers on network science topics. Google Scholar and arXiv are two good references for network science methods and application papers.

Download the example code files

You can download the example code files for this book from GitHub at https://github.com/ PacktPublishing/Modern-Graph-Theory-Algorithms-with-Python. If there's an update to the code, it will be updated in the GitHub repository.

We also have other code bundles from our rich catalog of books and videos available at https:// github.com/PacktPublishing/. Check them out!

Conventions used

There are a number of text conventions used throughout this book.

Code in text: Indicates code words in text, database table names, folder names, filenames, file extensions, pathnames, dummy URLs, user input, and Twitter handles. Here is an example: "This script shows that average subgraph centrality varies between the two subfamily trees, with Greenberg's average subgraph centrality of 2.478 and Dimmendaal's average subgraph centrality of 3.276."

A block of code is set as follows:

```
#compare subgraph centrality of language families
gs=nx.subgraph_centrality(G)
print(np.mean(np.array(list(gs.values()))))
gs2=nx.subgraph_centrality(G2)
print(np.mean(np.array(list(gs2.values()))))
```

Bold: Indicates a new term, an important word, or words that you see onscreen. For instance, words in menus or dialog boxes appear in **bold**. Here is an example: "When you hover over the **Movie DBMS** label on the right-hand side of the screen, you'll see a **Start** button that launches the connection to this database. Click on **Start**."

> **Tips or important notes**
> Appear like this.

Get in touch

Feedback from our readers is always welcome.

General feedback: If you have questions about any aspect of this book, email us at customercare@ packtpub.com and mention the book title in the subject of your message.

Errata: Although we have taken every care to ensure the accuracy of our content, mistakes do happen. If you have found a mistake in this book, we would be grateful if you would report this to us. Please visit www.packtpub.com/support/errata and fill in the form.

If you are interested in becoming an author: If there is a topic that you have expertise in and you are interested in either writing or contributing to a book, please visit authors.packtpub.com.

Share Your Thoughts

Once you've read *Modern Graph Theory Algorithms with Python*, we'd love to hear your thoughts! Scan the QR code below to go straight to the Amazon review page for this book and share your feedback.

https://packt.link/r/1-805-12789-6

Your review is important to us and the tech community and will help us make sure we're delivering excellent quality content.

Download a free PDF copy of this book

Thanks for purchasing this book!

Do you like to read on the go but are unable to carry your print books everywhere?

Is your eBook purchase not compatible with the device of your choice?

Don't worry, now with every Packt book you get a DRM-free PDF version of that book at no cost.

Read anywhere, any place, on any device. Search, copy, and paste code from your favorite technical books directly into your application.

The perks don't stop there, you can get exclusive access to discounts, newsletters, and great free content in your inbox daily

Follow these simple steps to get the benefits:

1. Scan the QR code or visit the link below

https://packt.link/free-ebook/9781805127895

2. Submit your proof of purchase
3. That's it! We'll send your free PDF and other benefits to your email directly

Part 1: Introduction to Graphs and Networks with Examples

This part of the book builds the practical and theoretical foundations of network science and introduces two Python packages useful in analyzing networks. *Part 1* details several examples of data science problems that can be formulated as network science problems, including problems to do with social relationship data, neural network architectures, ontologies, time series data, and spatiotemporal data. This part also establishes foundational topics in graph theory, including categories of graphs and formal definitions, and introduces the igraph and NetworkX Python packages through example networks.

Part 1 has the following chapters:

- *Chapter 1, What is a Network?*
- *Chapter 2, Wrangling Data into Networks with NetworkX and igraph*

1
What is a Network?

This chapter introduces the basics of graph theory and its applications in network science. Network science is not a commonly taught area of data science, but many problems can be framed through a network science perspective. Network-based algorithms often scale better than other machine learning algorithms, making them ideal for analyzing datasets with many variables, exploring spatial datasets with many locations represented, or spotting trends in high-dimensional time series data. Later chapters will delve more deeply into the topics with hands-on examples.

In this chapter, we will define terms that will be used throughout the book, explore some common uses of network science in analyzing social relationship data, and introduce two Python packages that will be used in subsequent chapters. After finishing this chapter, you'll start to recognize data science problems that can be formulated as network science problems and how to represent them visually in Python.

Formally, we will cover the following topics:

- Introduction to graph theory and networks
- Examples of real-world social networks
- Other type of networks

Technical requirements

We have very few technical requirements or assumptions for this chapter. If you have not installed Python, you are encouraged to do so, as we will be using Python regularly—specifically, the Jupyter Notebooks that are installed with the Anaconda version of Python installation. If you have difficulty with installation, support can be found on Stack Overflow. The code for each chapter can be found under this GitHub link: `https://github.com/PacktPublishing/Modern-Graph-Theory-Algorithms-with-Python`.

Introduction to graph theory and networks

Social connections are fundamental to society, including family ties, shared roles in communities, trading relationships, and many more. Social networks—from MySpace to Twitter to TikTok—have played a larger and larger role in marketing, job-hunting, and information sharing in the last decades. Software systems might even be built between engineers scattered across the globe collaborating on platforms such as GitHub or Slack to coordinate efforts from teams in Kenya, India, and Australia who have never met.

A branch of data science called **network science** studies these relationships between individuals, groups, and even societies within social networks using algorithms and statistical methods originating in a field of math called **graph theory**. Graph theory studies pairwise relationships between objects. Graph objects (including people, towns, ideas, points in time, and many more types of objects to study) are represented in network science as vertices, or points, within the network; note that some disciplines may use the term "*node*" instead of "*vertex.*" Relationships between objects (such as mutual collaborations between people or bridges connecting different islands or statistical connections between time points) are represented as edges within the network and connect pairs of vertices that share a relationship.

To make this a little more concrete, let's consider three young women: Ayanda in South Africa, Machiko in Japan, and Greta in Belgium:

Figure 1.1 – Ayanda, Machiko, and Greta

Perhaps Ayanda, Machiko, and Greta join an online women's coding hackathon aimed at sustainable energy solutions. Through the hackathon, they meet and form a team to work on solar power solutions for rural villages. They've formed a professional network with mutual relationships among the three women, as shown in *Figure 1.2*:

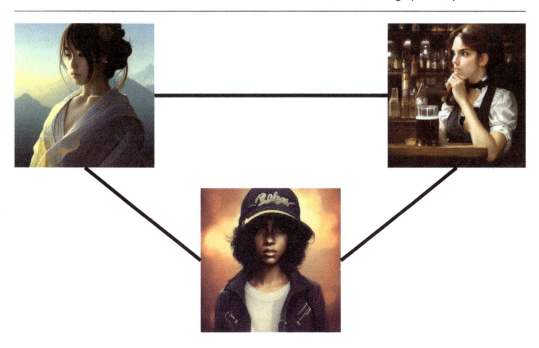

Figure 1.2 – A network representation of Ayanda, Machiko, and Greta
showing pairwise connections between the women

As these women work on their hackathon project, Ayanda may invite colleagues within her social network, adding to the team and creating new connections for Machiko and Greta over time. Perhaps Amara, a colleague of Ayanda's from Kenya, connects with Greta but not Machiko to code the backend of their project, adding another member to the network:

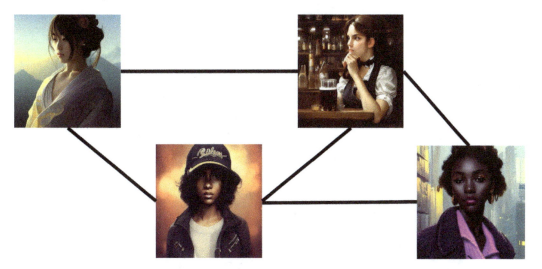

Figure 1.3 – Social network with the addition of Amara

The evolution of networks over time is a key problem in network science today, as we often have incomplete data on relationships that exist outside of what we explicitly observe and want to predict future phenomena, such as predicting the potential for bipartisan fake news spread in the next election or when new members may join a criminal organization based on ties that exist in the present. We will wrangle some problems related to dynamic social networks and their applications in *Chapter 8*, where we analyze the spread of disease over time as a social network evolves.

Before jumping into real-world problems and their solutions in Python, it's important to understand the fundamentals of network science. Let's dive into some of the terminology and conventions of graph theory and networks.

Formal definitions

Real-world networks are usually formalized and studied with tools from graph theory. As we've mentioned, graph theory defines relationships between objects in the form of vertices and the edges that connect related vertices. Let's formalize this a bit and explore the foundations of network science found in graph theory. A graph is defined as a set of vertices and a set of lines, or edges, that connect pairs of vertices (Berge, 2001). Thus, a graph represents the structure of a network well. Returning to our hackathon social network, we represent our colleagues as vertices and connect them with edges if they collaborate.

Graphs can represent different types of relationships. A self-relationship of a vertex can be shown with a loop connecting the vertex to itself. For instance, when writing emails, someone might copy themselves on the email or send an email to themselves as a reminder. Another type of relationship is directionality. For instance, reposting content is a unidirectional interaction (usually), but friending someone is a bidirectional interaction where both parties mutually connect. Unidirectional relationships are represented in a graph with directed edges, usually denoted with an arrow on visualizations of the graph. Bidirectional relationships have undirected edges, denoted as lines without arrows. *Figure 1.4* shows a graph with loops, directed edges, and undirected edges:

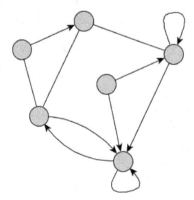

Figure 1.4 – A graph with loops, directed edges, and undirected edges

In scientific literature, the terms network and graph are used interchangeably, as are terms for edges and vertices. *Table 1.1* shows a few of the common terms encountered:

Network Science	Graph Theory
Network	Graph
Node	Vertex
Link	Edge
Collection	Set
Relationship	Function
Directed graph	Digraph

Table 1.1 – A terminology comparison between network science and graph theory

In practice, these distinctions are rarely made, so these two terminologies are often synonyms of each other. In this book, we'll stick with *edges* and *vertices* as our graph components and networks as our preferred term for the collection of vertices and edges created from real-world data.

Many types of graphs exist within graph theory that are realized in real-world data. Let's briefly define some types of graphs that exist:

- A directed graph contains one or more edges with a direction (to or from another vertex, called arcs)

- An undirected graph only has edges, which do not have an origin or destination vertex

"A graph is simple if it has no multiple lines (typically used to represent multiple relationships that exist or denote the strength of a relationship)." A simple undirected graph contains no loops, but a simple directed graph can contain loops. Let's visualize what these graphs look like in practice:

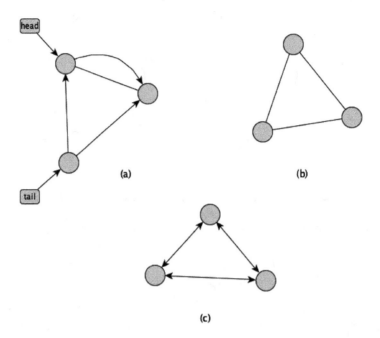

Figure 1.5 – (a) A directed graph, (b) simple graph, and (c) simple directed graph bottom

We can define vertices and their connecting edges or arcs based on data to create a network for visualization or analysis in Python. In the next section, we'll introduce some packages in Python that will be used throughout the book to create and study networks in practice.

Creating networks in Python

Now that we know a little bit about networks, let's explore how they are created with two Python packages commonly used in network science: igraph and NetworkX.

igraph is a network science software available in C, R, and Python. igraph provides many network science tools, including network creation options, many network analytics algorithms, and network visualization plots. Let's explore igraph network creation and visualization using our hackathon network:

1.   First, let's install igraph:

```
#install igraph and pycairo
!pip install igraph
!pip install pycairo
```

2. We'll then import `igraph` and its `Graph` module:

```
#import igraph
import igraph as ig
from igraph import Graph
```

3. Next, we'll use the `Graph` module to create a graph with three vertices with undirected edges connecting each vertex with the remaining vertices:

```
#create hackathon network
g_colleagues=ig.Graph(
    edges=[(0,1),(0,2),(1,2)],n=3,directed=False)
```

4. Now, we can use the `plot` function to visualize our hackathon network. igraph offers many customization options, including vertex color, vertex size, edge color, edge size, labeling of vertices, and the size of the plot (among others). For this plot, we'll bound the size of the image and create vertices large enough to label with the first initials of our hackathon network members:

```
#plot the hackathon network
ig.plot(g_colleagues,bbox= (200,200), vertex_size=40,
    vertex_label=["M","A","G"])
```

This should give you a plot that looks like *Figure 1.6*, showing *Machiko*, *Greta*, and *Ayanda's* hackathon network:

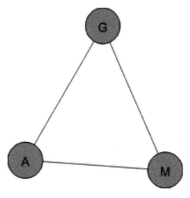

Figure 1.6 – A plot of the hackathon network in igraph

Now let's repeat our network creation in NetworkX. NetworkX contains much of the functionality of igraph but provides easier integration with other network science tools and platforms, as well as simple ways to explore differential equations on networks:

1. Let's first install `NetworkX`:

```
#install NetworkX
!pip install NetworkX
```

2. Now, let's import `NetworkX`:

```
#import NetworkX
import networkx as nx
```

3. NetworkX works a bit differently than igraph in the construction of a network. We first define an empty graph, fill in the vertices, and then define what edges exist:

```
#create hackathon network
G = nx.Graph()
G.add_nodes_from([1, 3])
G.add_edges_from([(1, 2), (1, 3), (2, 3)])
```

4. We'll need to add attributes (our network member initials), import `matplotlib`, and then plot our graph:

```
#plot the hackathon network
import matplotlib.pyplot as plt
G.nodes[1]['initial'] = 'M'
G.nodes[2]['initial'] = 'A'
G.nodes[3]['initial'] = 'G'
labels = nx.get_node_attributes(G, 'initial')
nx.draw(G, labels=labels, font_weight='bold')
```

This should show a plot similar in structure but a bit different in style to the one we constructed in igraph:

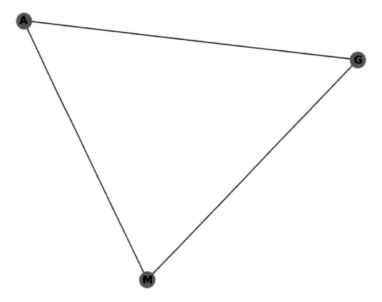

Figure 1.7 – A NetworkX plot of our hackathon network

We'll be using both igraph and NetworkX in the coming chapters to show examples in both packages. For some problems, igraph will have better functionality; for others, NetworkX is preferable. In practice, you'll encounter both as you implement network science on real-world problems. In the next sections, we'll overview different network types and problems, laying the foundation for our future chapters.

Random graphs

As network science began to tackle more and more real-world problems, understanding processes of network growth became an avenue of research:

Was edge-building purely random?

Were there limits to the number of edges that any vertex might cultivate given the number of vertices?

Did vertices with many edges tend to encourage edges among unconnected vertices that shared an edge with a particular well-connected vertex?

Did vertices and edges separate into well-connected groups that didn't connect well with other groups?

A few growth process models gave rise to common types of growth patterns that are useful in network science research. Some of these, such as **Erdös-Renyi networks**, serve as comparisons for organized development in real-world networks. Erdös-Renyi networks are random networks, where edges between vertices are created probabilistically between any pair of vertices. For instance, to create a network with many edges, we might set the probability of vertex pair connections to 70%. To create a network with few edges, we might set the probability of vertex pair connections to 10%. *Figure 1.8 (a)* shows an Erdös-Renyi network with edge connection probability of 70%; *Figure 1.8 (b)* shows an Erdös-Renyi network with edge connection probability of 10%:

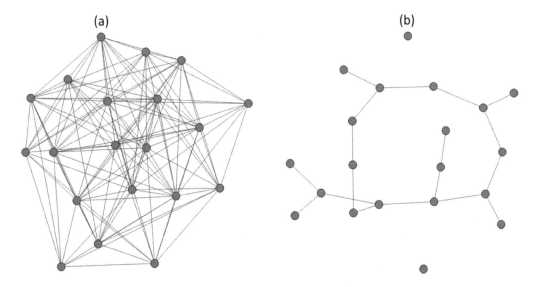

Figure 1.8 – Erdös-Renyi networks with edge probabilities of (a) 70% and (b) 10%

In the real world, preferential attachment, where some vertices attract connections to other vertices, is a more realistic model of how edges are added between vertices. One of the best-known examples of preferential attachment is the **Barabasi-Albert model**, where networks form more connections around certain vertices.

In a real-world example of preferential attachment, let's consider a frontend engineer searching social media for backend engineers who are potential collaborators. Thus, the frontend engineer's connections would not be random. They will be cultivated based on the engineer's specific needs and the results of a search for backend engineers. **Scale-free networks** are models of this preference for attachment that often happens in real-world network formation processes. *Figure 1.9* shows the frontend engineer's social network after adding the needed connections, who may not know each other before connecting with the frontend engineer:

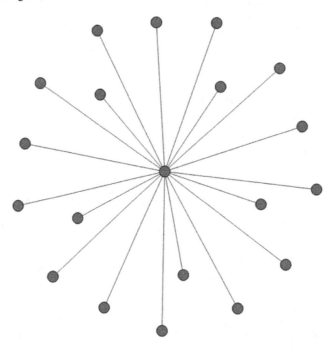

Figure 1.9 – A scale-free network of engineering connections cultivated
by the frontend engineer's search of social media

One of the major criticisms of scale-free models of network formation is the lack of hubs, or regions with high connectivity between a set of vertices and relatively few connections with other sets of vertices. **Watts-Strogatz models** account for this tendency to form denser regions of connections. As an example, gene networks often form clusters of related genes that influence each other functionally; some gene clusters also influence other gene clusters, forming occasional connections between gene sets with high connectivity, as shown in *Figure 1.10*:

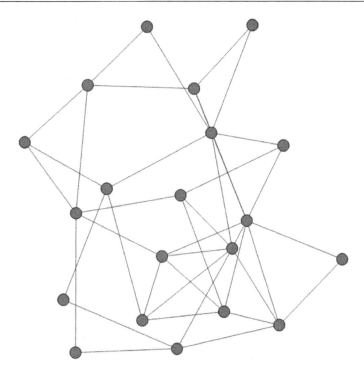

Figure 1.10 – A network of related genes

Recent research has shown that none of these random network models capture all the nuances that come with real networks. Many real networks have aspects of preferential attachment and aspects of dense clustering, as well as some random connections that would not be expected given what is known about the network. Friends of friends often meet. Individuals find others who share their interests by searching social networks or connecting to a friend's other social groups. Some friendships randomly start when individuals are stuck in an elevator or on a ferry.

Now that we have explored how to create networks in Python, let's turn back to some real-world examples of networks, including directed and undirected networks and networks that change structure over time.

Examples of real-world social networks

Real-world networks are often more complicated than the examples we've considered. Let's consider some nuances in social media networks. In the simplest forms, a social network contains mutual connections between friends at a single point in time. However, some friends may interact more often on the social media platform than others; they may comment on posts, reshare content, and message each other frequently. Weighted networks assign numerical values to edges in a network according to some measurement of relationship strength or frequency of interactions.

Let's return to our hackathon network and weight our edges based on the average number of conversations per day of the hackathon between different team members. Perhaps *Machiko* is coordinating the business use case of the technology and does not interact much with *Ayanda*, the backend engineer, but needs to make sure the front end is intuitive (*Greta's* team position). However, we'd expect *Ayanda* and *Greta* to work together more frequently, as they are integrating the backend and front end of the product. Perhaps we have a weighted network as shown in *Figure 1.11*:

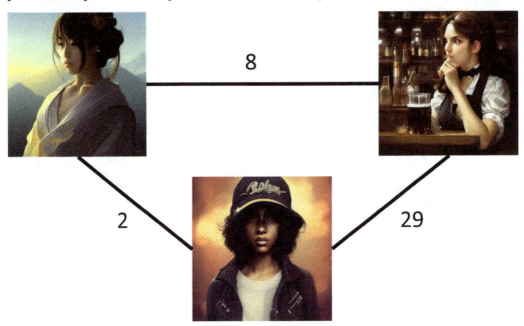

Figure 1.11 – A weighted hackathon network of conversations between team members

In addition to weighting, relationships can be undirected or directed. So far, we've considered undirected networks, where relationships are mutual. Friendships, Zoom calls, shared classes, team membership, and many other real-world networks involve mutual relationships. However, messages are one-way interactions, between a sender and receiver. Reposts of social media content are one-way relationships, as well. Gene regulation, population migration, airline flights, and goods transportation also involve one-way interactions. Directed graphs allow us to capture directional information in our graph, and like undirected graphs, directed graphs can be weighted or unweighted.

Let's consider a football game where the players pass a ball to each other over a 5-minute period. Perhaps one forward dribbles for a while and then passes the ball to the other forward, who quickly passes it to the open center striker. *Figure 1.12* shows this weighted network of football field interactions among the teammates:

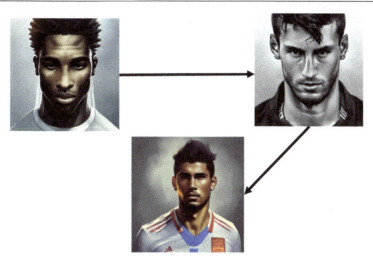

Figure 1.12 – A network of football teammates passing to each other over a 5-minute period

Networks are rarely static. Consider a group of friends graduating from university. Some will stay in touch over the next several years; some will lose contact. The strength of lasting friendships may change, as well. Those who stay in the same area where they went to university might strengthen their friendships as they enter the workforce. Those who move away may connect with others in their new location and start losing touch with friends in other areas. Let's consider four close friends graduating from university in Cape Town to pursue different careers in different cities:

Figure 1.13 – A friendship network of university graduates moving
to different cities and pursuing different careers

A decade later, some of these university friends are no longer in contact with each other, yielding a much different friendship network, shown in *Figure 1.14:*

Figure 1.14 – The university friendship network a decade after graduation

Figure 1.14 shows that one friend (in the upper-right corner) has lost touch with all but one close university friend. One friend (lower-left corner) has maintained all her close university friendships. This change may seem rather trivial, as all friends have maintained at least one friend from the original group. However, in terms of information exchange across the network, the original friendship network was more amenable to information exchange from the friend in the upper-right corner to the others in the network. She could communicate directly with each friend.

However, after a decade of drifting friendships, information exchange from the friend at the top right must go through the friend at the bottom left to reach the rest of the network. The time taken for important news (perhaps regarding a reunion or wedding) to reach the full network will be longer in the network in *Figure 1.14* than in *Figure 1.13*.

In this section, we've introduced a few types of social networks. We've seen how relationships can be mutual or one-sided. We've seen how networks can change over time and how this may impact information exchange on networks. We'll revisit social relationships in *Chapters 2, 3, 8,* and *9*. Now, let's explore some other types of problems that can be formulated as networks.

Other type of networks

So far, we've mainly considered social networks. However, objects and their relationships can be defined on a much broader range of problems. Network science provides useful tools for understanding spatial data. For instance, consider a region with several small island towns connected by a series of bridges that allow goods and people to travel from region to region. During an epidemic, it might be useful to know which routes to close to minimize the spread of disease from one town to another. However, cutting off routes entirely would leave some areas without necessary supplies, such as food or medicine. Tools we'll consider in later chapters can help regions plan the optimal routing of goods and minimize the potential spread of an epidemic.

We might also consider a map of stores, each with differing sale volumes of citrus fruit. Stores could be connected into a network based on shared management across stores, based on similarity of citrus fruit sales patterns, or based on geography (within the same county or country, for instance). How we define our network will impact what insight we can gain about regional citrus fruit sales and factors that might be impacting those sales patterns. We'll explore this use case more in *Chapter 4*.

Similarly, we can wrangle time series data into networks and use network science tools to gain insight into patterns over time. For example, consider daily financial data across the tech, agriculture, and manufacturing sectors. We can consider weekly correlations among the sectors and build a weighted network with edges defined by correlation values and track network properties over the weeks that exist in our data to pinpoint changes that might signal a coming market crash. Correlation networks among stocks and financial sectors are well studied, and we'll return to this example in *Chapter 6*.

Sometimes, we have spatiotemporal data, where our data has spatial and time series components. Consider the price of grain across several urban markets across a country. Pricing will vary across markets, with influences of local supply and demand, regulations in certain regions of the country, and local salaries likely influencing the price of grain at a specific market. Proximity to other markets and influences over time (such as the conflict in Ukraine limiting grain supplies globally) also influence pricing. By creating a series of networks based on geography, time period, and some of these other aforementioned factors, we can understand which markets tend to behave similarly with respect to grain price, which areas might be most vulnerable to price increases, and which areas might be most vulnerable to supply chain issues. This could allow an aid organization to pivot food aid more quickly to areas where people are unlikely to be able to afford future grain prices or may lose access to grain altogether. We'll tackle this problem in *Chapter 7*.

Now that we have seen some use cases in spatial and temporal data, let's turn to some more advanced problems in data science that can be formulated in terms of network science, including neural network architectures and language ontologies.

Advanced use cases of network science

The tools of network science extend beyond social networks, spatial data, and temporal data. Deep learning models are ubiquitous in data science today, solving problems in computer vision, natural language processing, time series forecasting, and generative artificial intelligence. **Large language models** (**LLMs**) and text-to-image generators rely on a type of deep learning architecture called transformer models, feed-forward neural networks that find patterns in data by embedding input data, tuning attention weights, and decoding the data with respect to the outcome. These models can have billions or even trillions of parameters to tune across many connected layers. When combined with pre-trained **contrastive language-image pre-training** (**CLIP**) models, transformer models such as DALL-E can even generate realistic images based on text input. For instance, inputting *hyperdetailed photorealistic king cobra, background desert market* into NightCafe's DALL-E algorithm produced the image in *Figure 1.15*:

Figure 1.15 – NightCafe's DALL-E output image for the prompt "hyperdetailed photorealistic king cobra, background desert market."

However, given the size of deep learning models such as transformer models (or the convolutional neural networks often used to classify images), understanding connections across layers is critical to building good initial models that won't require as much tuning and is helpful in tuning architectures based on network properties.

Additionally, networks themselves may be modeled through a special class of deep learning algorithms, called graph neural networks, that take networks as input data and build regression or classification models based on some output. For instance, say we want to understand which networks are most vulnerable to the spread of fake news; in the 2014 and 2018 Ebola outbreaks, fake news regarding public health measures and virus sources hindered public health efforts and resulted in a greater loss of life, as well as some violent attacks on treatment centers. We may have a few hundred networks that we have studied and classified as high, medium, or low risk based on network properties and simulations of fake news spread on these networks, but we want to assess several hundred thousand scraped real-world networks collected from social media sites around the world. We can train a graph neural network on our classified data and apply it to the much wider collection of scraped networks to classify their fake news risks.

Besides neural networks, many other advanced uses of networks exist, including in the mapping of ontologies. Ontologies organize relationships between objects, such as semantic relationships between words, regulatory relationships between genes, or symptoms shared between diseases. Often, multiple ontologies exist, such as the organization of consumer goods across store chains. Mapping one ontology to another provides a way to combine information from each ontology into a single entity. Given that each ontology exists as its own network being mapped onto other ontologies, the problem can be simplified to network mapping. We'll explore this in more depth later in the book.

Summary

In this chapter, we explored different types of networks, introduced two network packages in Python, examined different theoretical models of network growth and their limitations, and considered several examples of networks not based on social relationships. We also touched on some more advanced topics in network science today, including network applications to deep learning, network classification, and ontology mapping. In the next chapter, we'll dive deeper into igraph and NetworkX with practical network examples.

References

Aganze, E., Kusinza, R., & Bukavu, D. R. (2020). The current state of fake news in the DR Congo and social impacts. *Global Journal of Computer Science and Technology*.

Aric A. Hagberg, Daniel A. Schult and Pieter J. Swart "Exploring network structure, dynamics, and function using NetworkX". In Proceedings of the 7th Python in Science Conference (SciPy2008), Gäel Varoquaux, Travis Vaught, and Jarrod Millman (Eds), (Pasadena, CA USA), pp. 11–15, Aug 20

Berge, C. (2001, January 1). The Theory of Graphs. Courier Corporation.

BRANDES, U., ROBINS, G., McCRANIE, A., & WASSERMAN, S. (2013, April). What is network science? *Network Science, 1*(1), 1–15.

Csárdi G., Nepusz T. The igraph software package for complex network research. *InterJournal Complex Systems*, 1695, 2006.

Ducruet, C., & Beauguitte, L. (2014). Spatial science and network science: review and outcomes of a complex relationship. Networks and Spatial Economics, 14(3-4), 297-316.

Estrada E. *The Structure of Complex Networks*: Theory and Applications. OUP Oxford, 2012.

Fung, I. C. H., Fu, K. W., Chan, C. H., Chan, B. S. B., Cheung, C. N., Abraham, T., & Tse, Z. T. H. (2016). Social media's initial reaction to information and misinformation on Ebola, August 2014: facts and rumors. Public health reports, 131(3), 461-473.

Hagberg, A., & Conway, D. (2020). Networkx: Network analysis with Python. `https://github.com/networkx`.

Kim, M., & Sayama, H. (2017). Predicting stock market movements using network science: An information theoretic approach. Applied network science, 2(1), 1-14.

Kiss, I. Z., Miller, J. C., & Simon, P. L. (2017). Mathematics of epidemics on networks. Cham: Springer, 598, 31.

Mehler, A., Lücking, A., Banisch, S., Blanchard, P., & Job, B. (Eds.). (2016). Towards a theoretical framework for analyzing complex linguistic networks (pp. 3-26). Berlin: Springer.

Mocanu, D. C., Mocanu, E., Stone, P., Nguyen, P. H., Gibescu, M., & Liotta, A. (2018). Scalable training of artificial neural networks with adaptive sparse connectivity inspired by network science. Nature communications, 9(1), 2383.

Valente, T. W. (2005). Network models and methods for studying the diffusion of innovations. Models and methods in social network analysis, 28, 98-116.

2

Wrangling Data into Networks with NetworkX and igraph

In this chapter, we will introduce many types of data that are common in analytics projects, including **spatial data**, **temporal data**, and **spatiotemporal data**. Because these data structures do not exist as networks, they must be preprocessed into networks for further analytics. In future chapters, we will detail strategies for optimal preprocessing, but the goal of this chapter is to get familiar with the basics and how preprocessing works with the NetworkX and igraph packages.

We'll consider many real-world problems in this chapter and the remaining chapters to build intuition around data that can be reformatted and analyzed as a network science problem. Oftentimes, network-based algorithms have lower computational costs than algorithms designed for time series analytics or spatial data. By the end of this chapter, you'll be able to recognize many types of problems that work well with network analytics, and you'll walk away equipped to dive into the problems in later chapters.

Specifically, we will cover the following topics in this chapter:

- Introduction to different data sources
- Wrangling data into networks with igraph
- Social network examples with NetworkX

Technical requirements

To run the practical examples in this chapter, you need to be familiar with Python programming and must install the igraph and NetworkX Python packages.

The code for this chapter is available here: https://github.com/PacktPublishing/Modern-Graph-Theory-Algorithms-with-Python

Introduction to different data sources

In practice, we'll rarely create a network from scratch or come across a data source that naturally occurs as a network. We often must create a network from different data sources, including survey or sensor data, geographic data, time series data, demographic data, or even output data from machine learning models. In this section, we'll overview some common data sources before diving into two practical examples of data wrangling with igraph and NetworkX. Let's dive into our first data source, social interaction data.

Social interaction data

Much of network science originated with social networks, which capture relationships or interactions between individuals. Marketing campaigns often recruit individuals with many connections to others within a demographic group of interest, as well as those whose connections also have many connections within a demographic group. For instance, a toy manufacturer might have a new toy coming out for the Christmas season, perhaps a *steampunk nutcracker ballet mouse king doll* (shown in *Figure 2.1*), that they would like to market to European girls between the ages of 5 and 11 based on market research about which age and gender groups are most likely to see the Nutcracker ballet:

Figure 2.1 – Marketing advertisement for the new Mouse King doll debuting at Christmas

The toy manufacturer may look for social media influencers (individuals with many subscribers on YouTube or followers on TikTok) whose network of viewers includes mostly girls between 5 and 11 years old who are viewing the influencer's content from Europe. Sometimes, this data is readily available by scraping influencers' content. Other times, this data needs to be collected in pieces and analyzed first to narrow down influencers among the demographic groups of interest through content curation and the scraping of related content metrics.

Another common source of data for social network construction is **survey data**, in which individuals rank friendships or interaction patterns with others in a social group. For instance, students might indicate other students with whom they study, socialize, or attend classes. Students who interact with each other in many settings share stronger social interactions, while students who interact infrequently or only in certain settings share weaker social interactions. Weighted social networks can visualize not only interactions that exist but also their strengths. In social science settings, stronger ties often play an important role in the transmission of information or the adoption of products or behaviors. For instance, when an adolescent begins smoking or drinking alcohol, individuals with strong ties to that individual are at higher risk of starting those behaviors themselves. Within highly connected groups of adolescents, the risk of behavior spreading is high among the group when one individual adopts a risky behavior.

Weaker social ties can be just as important as strong social ties (or even more important!), and within the field of sociology, these ties form the basis of an individual's social capital. Social capital measures the resources upon which an individual can draw when needed from those with whom the individual has connected or those connected to an individual's connections. Within the context of job searches, a job seeker's strongest and most-lasting ties may not be what is needed to find a new job. However, it is likely that friends-of-friends and acquaintances form a much larger social network, and these ties may present new employment opportunities to the job seeker. Larger networks of friends-of-friends and casual acquaintances often provide more opportunities to find that next job through word-of-mouth or job ad posts on social media.

We'll return to survey-based social network data in this chapter's *Wrangling data into networks with igraph* section where we construct a network of students based on endorsement interactions across social contexts.

Spatial data

Another common source of network data is data collected from geographic information systems or organized by geography. **Spatial data analytics** is a branch of data science that analyzes geography-based data; however, when data covers large geographic regions or involves complex statistical calculations, it can be easier to formulate the data and problem in terms of network science. Network algorithms are often faster than spatial data algorithms, and it is easy to store large networks in graph databases such as Neo4j.

Consider the worldwide spread of COVID-19 through travel networks. The pandemic originated in China but quickly spread to other parts of Asia, as well as the rest of the world, through the travel of infected but not terribly sick individuals on airplanes, trains, and other means of transportation (such as flights from Beijing to Paris, as shown in *Figure 2.2*). Air travel also allowed new variants to spread from one continent to another during the pandemic. Fortunately, COVID-19 had a low case fatality rate and did not cause severe disease in most people who were infected. However, a highly infectious disease similar to COVID-19, but with a high fatality rate and a long incubation period from infection to the display of symptoms (such as a more infectious version of Ebola) spreading through the same travel networks would create a very severe global health crisis before it was detected. As we'll see in later chapters, epidemic threats with spatial and social components can be modeled and simulated through network science.

Beijing to Paris flight

Figure 2.2 – A flight from Beijing to Paris during the initial transmission period of COVID-19

Health-related networks aren't the only spatial data sources useful in data science. One common source of spatial data is retail data, where stores exist in different locations and may carry different products, sell products at different rates, and adjust prices to match regional norms. Customer demographics may also vary significantly across locations and produce very different buying patterns. For instance, consider customers' buying behavior at a store near a university versus customers' buying behavior at a store in the suburbs. The store near a university may have more late-night purchases, sell a lot of snacks, and rarely need to stock diapers relative to the store in the suburbs. Analysis of customers' buying behavior, goods pricing, or out-of-stock frequency across products and geographies of stores can yield valuable strategic insight for individual stores and their parent chains interested in optimizing stocks, prices, and timing of promotions. We'll wrangle data related to millet prices at local markets into a network later in this chapter.

Temporal data

Networks are rarely static. Social networks evolve over time as connections forge and break. Sales patterns change as new items are added, needs for specific items change over seasons, or changes in the economy spur changes in consumer behavior. Stock market trading volumes and industry/regional correlations change over time as political and business climates change.

All these changes involve a time component in the data. There are a few ways to model time trends in networks, and we'll start with two use cases and build out more uses in *Part 3* of this book.

First, let's return to our example of a job seeker leveraging his or her social network to find a new job. As this individual begins a job search, they may add suggested professional connections on LinkedIn, attend local networking events, or connect with a few recruiters looking to hire; all these actions add connections to their social network (and increase the likelihood of finding a new job). *Figure 2.3* shows women meeting at a networking event, forging new professional connections to leverage in their job searches:

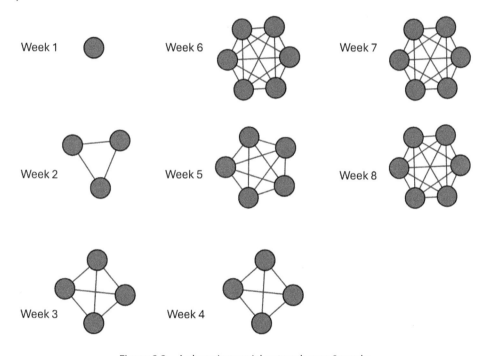

Figure 2.3 – A changing social network over 8 weeks

We can represent changes over time as a series of networks, each denoting a specific time when we capture the job seeker's social network. We may collect the data daily or weekly or even monthly for the period we are interested in analyzing. Let's say we capture data weekly for two months. This will give us roughly eight social networks, which we can analyze and summarize to track changes in network size and density of connections over time. Perhaps we also have data related to interviews our job seeker obtained, which we can examine along with the changes in their social network to gain insight into how social network growth has helped our job seeker obtain a new job.

To study how social ties and specific activities increase the likelihood of job interviews and offers, we could recruit a few hundred job seekers, collect information on interviews/offers and their current social network each week for eight weeks, and build a regression model to predict job interviews or job offers based on prior weeks' social network metrics. In a more sophisticated analysis, we could use a type of regression model that includes time series components to account for time effects, as well as social network metrics and initial data on job seekers (such as demographic data, career information, educational history…). We'll consider this type of model in *Chapter 13*.

Spatial data can also include time components, where new flights are added between countries or where stores' management structure changes over time to create different connections between stores within a retail chain. We've already seen how we can wrangle spatial data into networks. For **spatiotemporal data**, we can build networks at an instant in time where we sample the data (as in our job networking example), or we can chunk our data into time periods, create networks for each time period, and then consider the full series of networks in our analysis. In the retail store example, we might have real-time purchase behavior to parse into weekly trend networks, where stores are connected based on the similarity of sales across the full inventory of store pairs. We might be interested in analyzing how buying trends change over the month of December, when people celebrate New Year's, Christmas, Hanukkah, Kwanzaa, and end-of-year business deals. We'll consider examples of this in the *Wrangling data into networks with igraph* section.

Biological networks

Network science plays a prominent role in modern genomics research. Environmental factors, such as stress or eating habits, can modify which genes are expressed within a cell or tissue by making transcription of **deoxyribonucleic acid (DNA)** more or less likely for a set of genes, leading to higher or lower levels of proteins produced by translating the genetic code to **ribonucleic acid (RNA)** and then to proteins (*Figure 2.4*):

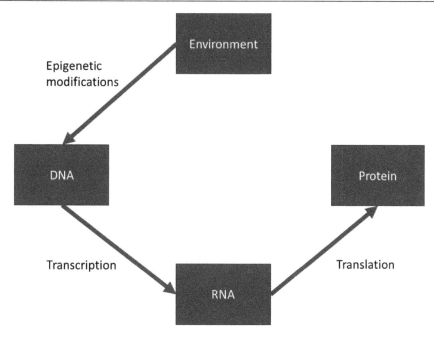

Figure 2.4 – A diagram showing the processes by which genes are
expressed, transcribed, and translated into proteins

Let's consider a concrete example. Cobras are venomous snakes, and their DNA encodes genes related to venom production. Transcription of these genes happens in the venom glands that produce venom and store it until the snake strikes a target. The DNA encoding these genes is transcribed into RNA, which is then translated into a protein.

Gene sets transcribed together are said to be **coexpressed**; **coexpression networks** are useful in understanding the relationship between genetic factors and diseases or physical traits. In our snake venom example, genes encoding venom would be coexpressed in the venom gland, as all of the venom is produced in that part of the snake's body. Environmental factors can indirectly influence the expression patterns of genes by modifying the DNA structures to upregulate or downregulate transcription. For instance, when our snake bites in self-defense, it may empty its venom gland completely with the bite. This will cause DNA related to venom genes to be upregulated, allowing for more venom production to replenish the snake's supply so that it can eat and defend itself.

The field of **epigenetics** studies these gene-environment interactions, particularly in the development of complex diseases such as cancer or mental health disorders. Typically, gene expression within or across tissues is measured through protein binding on genes of interest using some sort of microarray chip. Many human diseases involve a more complex interplay of environment and genes than the regulation of venom levels in a snake's venom gland; some of these diseases involve many parts of the body.

For example, some people face higher risks of alcohol use disorders based on their aggregation of genetic risk factors; when a person at risk of alcohol use disorder begins drinking, alcohol crosses the blood-brain barrier and produces epigenetic changes in gene expression, including changes in expression of the genes that put the person at risk for the disorder. Microarrays measure the level of gene expression experimentally. Microarray studies of coexpression of genes in those with alcohol use disorders and those without alcohol use disorders provide insight into genetic risk and the epigenetic differences that occur because of alcohol consumption. Networks provide a convenient way to summarize expression similarity relationships and can be mined for subnetworks related to modules of related genes that perform a similar task (such as producing neurotransmitter binding sites related to serotonin or dopamine binding at synapses).

A related phenomenon in genetic studies is the regulation of gene expression, typically through epigenetic pathways, as well. As we've discussed, external environmental factors can influence gene expression; often, proteins created from gene transcription and translation regulate their own transcription, turning expression up or down based on the levels of proteins that exist. For instance, disruptions in gene regulation related to serotonin receptors in the frontal cortex are thought to underlie major depressive disorder, and disruptions to downregulation of cell growth genes are thought to underlie a lot of common cancer pathways, leading to uncontrolled growth and mitosis (tumors).

Directed networks provide an ideal way to summarize gene regulation pathways. Different biological and environmental mechanisms upregulate or downregulate the genes of interest. Genes may cluster into similar regulatory patterns within the mechanisms of study (protein expression, external exposures in the environment…). This gives insight into specific pathologies of interest or the impacts of environmental factors like stress.

Genomic studies sometimes leverage network science to map gene ontologies (networks that summarize hierarchical relationships of individual genes or genetic modules) to disease ontologies (networks that summarize categories of disorders). We'll dive deeper into ontology research in *Chapter 11*, but let's review some basics here. The mapping between ontologies allows researchers to formulate hypotheses regarding unknown genetic links to disorders based on known links between genes and similar diseases. Returning to substance abuse disorders, many genetic risk factors are known and shared across addictions; it is likely that newer behavioral addictions, such as video game addiction, share at least some of these genetic factors with other addictive disorders. Ontology mapping between genes and diseases provides a starting point for studying these new disorders.

Genomics is one of many types of data where network solutions aren't an obvious toolset for solving data problems. Let's look at some other examples where network science has provided insight not possible with other analytics tools.

Other types of data

Many other sources of data exist, and most can be wrangled into network form. Let's return to our discussion of **deep learning**. In the training process of deep learning, nodes in each layer of the neural network forge and break connections with nodes in other layers based on fit metrics in each iteration of the training. Let's consider a training process with four training epochs of a very small feedforward neural network with an input layer of two nodes, a hidden layer of three nodes, and an output layer of two nodes that originates as a fully-connected network, undergoes two epochs of pruning, and then adds back one of the pruned connections between the hidden and output layers, shown in *Figure 2.5*:

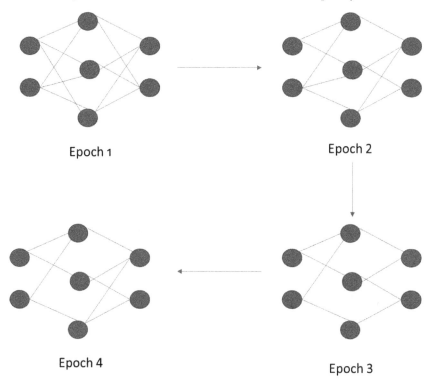

Figure 2.5 – The evolution of a simple deep learning model across training epochs

In *Figure 2.5*, we can see how each epoch has a different network structure across the deep learning model. Typically, we calculate fit statistics for each epoch, and perhaps we are interested in how properties of the deep learning network structure influence training accuracy on training and test sets across epochs. While this model is quite simple, deep learning models in practice may have several layers, directed edges between vertices (nodes in deep learning parlance), and many epochs of training. This creates a rich set of networks and network metrics across training to relate to model accuracy, which may be of use when tuning the model or changing initial parameters (perhaps adjusting pruning parameters when fine-tuning a pre-trained transformer model).

Fine-tuning pre-trained models is a common approach in *computer vision* and *large language modeling*. New training data to cover domain-specific use cases may be added, and model parameters need adjusting to incorporate the new training data into the full model. For instance, a large language model housed on HuggingFace might work well for embedding most scientific papers for cluster analysis but require fine-tuning to work well when plasma physics papers are included in the sample of scientific papers. Understanding the network structure of the original large language model and its training evolution can help researchers set optimal parameters in the fine-tuning process.

Another example of data that can be wrangled into network form to study is **linguistics data**. Many languages that exist today evolved from much older languages. For instance, the Semitic languages that exist today, including Hebrew and Arabic, originated from a Proto-Semitic language, which split into different regional branches as proto-Semitic speakers migrated and settled across the Middle East. One language, East Semitic (Akkadian) did not further branch to other languages; the other language, West Semitic, broke into South Semitic and Central South Semitic (which further broke into new languages over time). *Figure 2.6* shows the initial evolution of Proto-Semitic into new language groups:

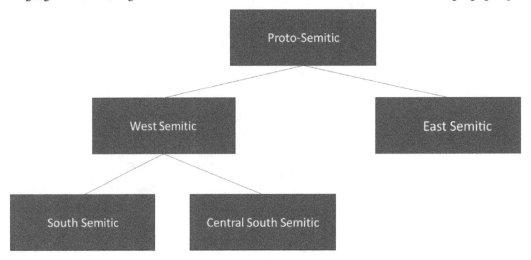

Figure 2.6 – A diagram of early language evolution from Proto-Semitic

Network science can help us compare language evolutions from original proto-languages to modern languages through the analysis of language family hierarchies. Comparing the structure of language evolution can illuminate ancient migration and trading patterns that can be difficult to discern from ancient records or archaeological sites. In addition, this comparison allows us to forecast future language splits based on the past behavior of the language family, similar evolution patterns seen in other language families, sociolinguistic data on language usage patterns, and the usage extent of current pidgins or dialects. We'll return to linguistic applications in *Chapter 11*, where we consider language ontologies of languages spoken in Africa and their evolutions over time.

For now, let's turn our attention to network creation in igraph and NetworkX from raw data files. This will equip us to create networks to analyze with tools we'll introduce in the coming chapters. First, we'll consider an example involving spatiotemporal data: quarterly millet prices across provincial markets in Burkina Faso.

Wrangling data into networks with igraph

Understanding trends in goods pricing is critical in many applications. For instance, a chain of gas stations may want to understand the differences in pricing across its locations. Often, local laws require gas stations within a certain distance of each other to be priced within a certain range of each other, creating local distributions of pricing across chains and within chains. Understanding this can help optimize prices regionally given constraints in each area or tie the data back to regional sales data to see how price distributions impact sales.

Consumers can also benefit from analyses of goods pricing across geographic areas. Consider local food markets in the developing world. Prices are rarely fixed within a country or region. They are dictated by supply, demand, and cost to the merchant selling the goods. Grains may need to be imported from other countries, whose prices will fluctuate with economic and climate conditions. Fruit prices may vary with season and local growing conditions. Clothing may increase in price as dyes become scarcer in the local area. For instance, consider a grocery store, shown in *Figure 2.7*, with supplier contracts and the financial backing of a parent chain. It may experience fluctuations in goods availability or price but will likely have the same goods at prices that don't fluctuate a lot:

Figure 2.7 – An illustration of a grocery store, part of a larger chain

Contrast this with a local market comprised of different vendors who may or may not be at the market on a given day and whose goods' availability and prices might fluctuate on the days they sell at the market. Larger markets, such as the one shown in *Figure 2.8*, with more established vendors and peak hours might show less variation than the more informal markets that line the streets throughout the day.

Figure 2.8 – An illustration of a large, formal marketplace in Burkina Faso, where
there is some commonality from day to day in vendors and their goods

When considering geography, we often want to consider not only the variable(s) of interest but also how close together the locations are relative to each other. Prices, sales, and goods availability in regions near each other likely reflect local conditions that aren't observed in the dataset, such as local weather, the local political context, and the number of competing local merchants. We'll discuss this topic further in *Chapters 3*, *4*, and *5*, but for now, it suffices to know that we can assign weights to different locations according to their relative distances.

Another factor to consider when wrangling spatial data is the metric of choice. Many spatial statistical metrics exist, and those of you familiar with spatial data analytics may want to use more sophisticated metrics than the ones we'll consider in this text. The **local Moran statistic**, a common metric in spatial data analytics, sums the weights between spatial areas and a function defined on the data in a pairwise manner; here, we construct the statistic using weight matrices we define and the correlation function, which should be familiar to readers as a summary statistic of likeness between data points.

Let's dive into our example dataset, derived from Humanitarian Data Exchange's Burkina Faso – Food Prices dataset from the World Food Programme Price Database (https://data.humdata.org/dataset/wfp-food-prices-for-burkina-faso; accessed July 15, 2022). In this dataset, we have pre-selected millet prices across 45 Burkina Faso province markets and aggregated data by quarter, starting in Quarter 2 of 2015 and ending in Quarter 2 of 2022. This period covers the COVID-19 pandemic and the beginning of the Ukraine War, both of which influenced supply chains. Given that markets in neighboring provinces may be impacted by the same local factors and supply chains, we have created a spatial weight matrix of 1's and 0's, representing adjacent provinces and non-adjacent provinces respectively.

To run all of the code, it may be necessary to install other packages that are dependencies of igraph or NetworkX, such as pycairo for visualizations. Each machine may function differently or have other packages already installed if you are using a cloud platform; if you run into any difficulties, please consult the igraph or NetworkX help guides' suggestions for running the packages on your operating system.

Let's first import the needed packages to construct our network in igraph with Script 2.1:

```
#import packages
import igraph as ig
from igraph import Graph
import numpy as np
import pandas as pd
import os
```

When importing data, you'll need to specify your own file path after you download the dataset. We import both the pricing data and the weight matrix from a local machine by adding to Script 2.1:

```
#import Burkina Faso market millet prices from the csv file
File="C:/users/njfar/OneDrive/Desktop/BF_Millet.csv"
pwd=os.getcwd()
os.chdir(os.path.dirname(File))
mydata = pd.read_csv(os.path.basename(File),encoding='latin1')

#import weight matrix of Burkina Faso markets
File="C:/users/njfar/OneDrive/Desktop/weights_bk.csv"
pwd=os.getcwd()
os.chdir(os.path.dirname(File))
weights = pd.read_csv(os.path.basename(File),encoding='latin1')
```

Now that we have our data imported, we can construct our local Moran statistic by computing the correlation coefficient across markets (ignoring the time indicator column) and multiplying the result by the weight matrix:

```
#define the metric between markets and construct the local Moran
#statistic
#here, correlation coefficient
data=mydata.iloc[:,1:46]
weights_total=weights.iloc[:,1:46]
cor=np.corrcoef(data.transpose())
cor[cor>0]=1
cor_weighted=np.multiply(cor,weights_total)
```

Now that we have our local Moran statistic constructed as `cor_weighted`, we can turn our data into a network with igraph. Our graph does not have any directionality of relationship, so we can set the mode to `undirected`. However, because our correlations include autocorrelations within individual markets, we need to remove loops from our network. We'll do this by defining possible self-loops and removing them from our edge list:

```
# create market graph and get rid of loops created from the correlation
bf_market_w=Graph.Adjacency(cor_weighted,mode="undirected")
edge_list=bf_market_w.get_edgelist()
self_loop=[]
for i in range(0,46):
    self=(i,i)
    self_loop.append(self)
to_remove=[]
for i in edge_list:
    for j in self_loop:
        if i==j:
            to_remove.append(i)
bf_market_w.delete_edges(to_remove)
```

Now, we can visualize our network to see which markets are isolated by plotting our results:

```
#create plot
ig.plot(bf_market_w)
```

This should give you a plot that shows one isolated market and several regionally-connected markets:

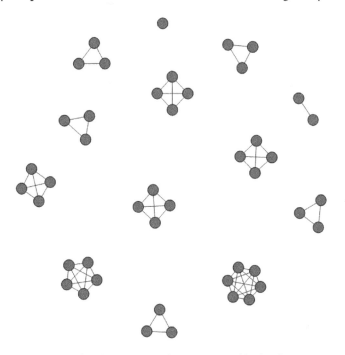

Figure 2.9 – A plot of Burkina Faso markets connected by local Moran statistics

Figure 2.9 shows that many markets form regional clusters, where prices are correlated within adjacent provinces. One market is isolated. Most markets connect into groups of three or four markets. One region contains six markets that are regionally connected and connected with pricing trends. Likely, this network visualizes regional connectivity rather than a mix of price correlation and region. We can visualize highly correlated regions by applying a threshold to the correlation calculations before computing the local Moran statistic.

Let's set the correlation threshold to 0.9 (very high correlation of prices) by modifying Script 2.1 to include thresholding in the local Moran statistic calculation piece:

```
#define the metric between markets and construct
#the local Moran statistic here, correlation coefficient
data=mydata.iloc[:,1:46]
weights_total=weights.iloc[:,1:46]
cor=np.corrcoef(data.transpose())
cor[cor>=0.9]=1
cor[cor<0.9]=0
cor_weighted=np.multiply(cor,weights_total)
```

```
# create market graph and get rid of loops created from the
#correlation
bf_market_w=Graph.Adjacency(cor_weighted,mode="undirected")
edge_list=bf_market_w.get_edgelist()
self_loop=[]
for i in range(0,46):
    self=(i,i)
    self_loop.append(self)
to_remove=[]
for i in edge_list:
    for j in self_loop:
        if i==j:
            to_remove.append(i)
bf_market_w.delete_edges(to_remove)

#create plot
ig.plot(bf_market_w)
```

This should yield many more isolated markets that are not highly correlated over our time period:

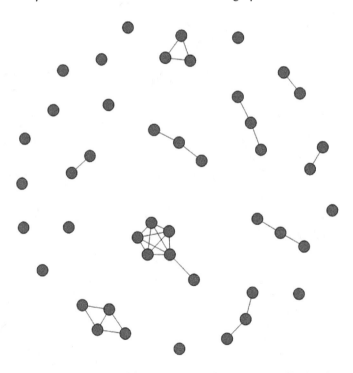

Figure 2.10 – A plot of Burkina Faso markets connected by local
Moran statistics over the selected threshold of 0.9

Figure 2.10 shows a much sparser network, where many markets are isolated. Some markets that were regionally connected still show connections among markets; however, many are not mutually connected anymore. Interestingly, our group of six regional markets is still connected with most connections intact after the thresholding. Indeed, only one market in that region shows a difference in connectivity relative to our non-thresholded version.

While this example considers correlations across the entire time period, it is possible to slice spatiotemporal data into overlapping time slices, create time-aware metrics like the local Moran statistic (with or without thresholding), and visualize the metrics' changes over time by plotting a series of networks created from these time slices and their network-defining metrics. We'll explore this notion further in *Chapter 7* when we break our Burkina Faso millet dataset into time slices to analyze changing graph metrics over time and space.

Now that we know how to wrangle data into an igraph network, let's see how this is done in NetworkX with another example.

Social network examples with NetworkX

Here, we'll consider a social network of **African Institute of Mathematical Sciences (AIMS)** students collected from a 2022 cohort in Cameroon. Students in this program live together in the same building, interact during class, eat meals together, and study together. However, students who share a country, come from the same undergraduate university, and speak the same language tend to interact more frequently than those who come from different backgrounds. Interactions such as those shown in *Figure 2.11* thus happen either organically as a function of background or artificially through shared programs and living arrangements:

Figure 2.11 – An illustration of a classroom setting, where students share courses within a program

To understand these interactions within our cohort of students, we administered a questionnaire about interactions with other students in various settings. The data was collected between November and December of 2022. The following are some example questions:

- Whom do you like to sit with during lunch or dinner?

- Who do you like to ask questions when you are having difficulty with a subject or course?

From these questions, students' perceptions of their networks could be examined to see the degree to which they agree. Upon looking at the data, we realized most interactions were reciprocated. If *Student A* endorsed survey items of closeness to *Student B*, *Student B* was likely to endorse those items, as well. Thus, the data form an undirected network of mutual interactions.

The dataset we'll use includes several fields. In the first column, we have the name of the surveyed student. In the next four fields, we have demographic factors (to which we'll return in later chapters), including country of origin, age, field of study, and undergraduate background. In the remaining five columns, we have five friends that the student most strongly endorsed.

Let's start by importing the packages we'll need to wrangle this dataset into a network with `Script 2.2`:

```
#import needed packages
import pandas as pd
import networkx as nx
import matplotlib.pyplot as plt
import numpy as np
import math
```

Now, let's import the dataset:

```
#reading the dataset
fichier = pd.read_csv("C:/users/njfar/OneDrive/Desktop/AIMS_data.csv")
data = pd.DataFrame(fichier)
```

To turn this data into a social network with NetworkX, we'll need to populate our network with vertices and edges, which we can do through loops. We'll first add vertices that include metadata including students' names and demographic characteristics. Then, we'll add edges that connect vertices based on the five closest relationships a particular student endorsed in our survey by adding to Script 2.2:

```
#intializing the social network
aimsNetwork = nx.Graph()
#populating the network with nodes and edges
for i in range(len(data["Name"])) :
    aimsNetwork.add_node(data["Name"][i],
        Age = data["Age"][i], country=data["Country"][i],
        Field=data["Field"][i],
        background=data["Background"][i])
    for j in range(len(data["Name"])) :
        aimsNetwork.add_edge(data["Name"][j],data["Friend 1"][j])
        aimsNetwork.add_edge(data["Name"][j],data["Friend 2"][j])
        aimsNetwork.add_edge(data["Name"][j],data["Friend 3"][j])
        aimsNetwork.add_edge(data["Name"][j],data["Friend 4"][j])
        aimsNetwork.add_edge(data["Name"][j],data["Friend 5"][j])
```

This gives us our initial network of AIMS students and their closest friends within the program. Let's plot this data to visualize our AIMS student social network. Because we want to show the names of students in our plot, we'll first compute the degree of each vertex, a centrality metric that we'll revisit in later chapters. Here, it suffices to show the importance of different students to the network and allow vertices to be large enough to visualize the names of students. Let's do this by adding to Script 2.2:

```
#plot AIMS student social network
Degree=aimsNetwork.degree()
var = [500*k[1] for k in list(Degree)]
plt.figure(figsize=(20,20))
nx.draw_spring(aimsNetwork, font_size=10, node_size = var,
    with_labels=True, node_color="red")
plt.show()
```

This script should show an image similar to *Figure 2.12*, which plots the social connections between AIMS students:

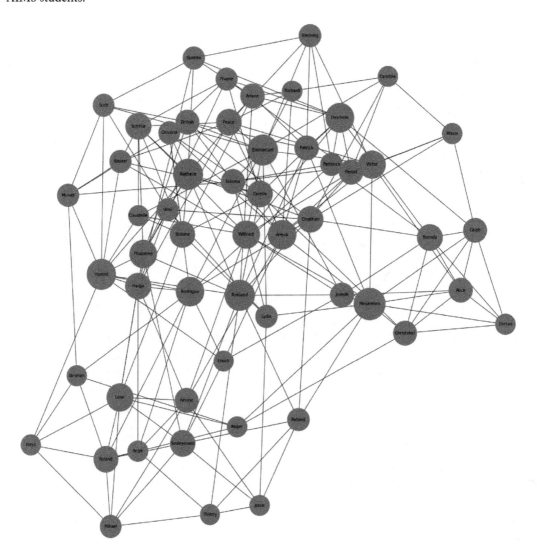

Figure 2.12 – Plot of AIMS Cameroon 2022 students' social interaction network

Figure 2.12 shows the network of AIMS students we created from our survey data. Note that some students only have the five closest friends as connections while others have many more close friends. This is typical of a social network, even one created from survey data. Some individuals have more connections than others. Some connect to different parts of a network, linking separated communities. Some highly connected individuals connect to other highly connected individuals, forming dense regions within the larger network. These properties play an important role in network analytics, and we'll dive deeper into their meanings and computation in later chapters.

Summary

In this chapter, we've explored use cases of networks in more depth, wrangled a spatial market dataset into a network based on regional connectivity and millet price correlation in igraph, and constructed a survey-based social network in NetworkX. Now that we have the basic tools needed to construct networks in Python from real data sources, in the next chapter, we can turn our attention to some real-world applications of network science, where we not only construct a network but analyze it for insights that solve important problems in science today.

References

Berger, S. L., Kouzarides, T., Shiekhattar, R., & Shilatifard, A. (2009). An Operational Definition Of Epigenetics. *Genes & development, 23*(7), 781-783.

Caballero, J. (2015). Banking crises and financial integration: Insights from networks science. Journal of International Financial Markets, Institutions and Money, 34, 127-146.

Christakis, N. A., & Fowler, J. H. (2013). Social contagion theory: examining dynamic social networks and human behavior. Statistics in medicine, 32(4), 556-577.

Dubos, R. (2017). Social capital: Theory and research. Routledge.

Ducruet, C., & Beauguitte, L. (2014). Spatial science and network science: review and outcomes of a complex relationship. Networks and Spatial Economics, 14(3-4), 297-316.

Kiss, I. Z., Miller, J. C., & Simon, P. L. (2017). Mathematics of epidemics on networks. Cham: Springer, 598, 31.

Krishnan, H. R., Sakharkar, A. J., Teppen, T. L., Berkel, T. D., & Pandey, S. C. (2014). The epigenetic landscape of alcoholism. International Review of Neurobiology, 115, 75-116.

Moyano, L. G. (2017). Learning network representations. The European Physical Journal Special Topics, 226(3), 499-518.

Sharma, S., Kelly, T. K., & Jones, P. A. (2010). Epigenetics in cancer. Carcinogenesis, 31(1), 27-36.

Sorrells, T. R., & Johnson, A. D. (2015). Making sense of transcription networks. Cell, 161(4), 714-723.

Part 2:
Spatial Data Applications

Part 2 builds on *Part 1* by introducing spatial data applications, including examples of trend spread across African countries, epidemic spread across student networks, transportation logistics problems involving roadways and locations of interest, snake conservation across regions of a national park, and text-based city park evaluation clustering. This part introduces epidemic (and more general differential equation) models, shortest path algorithms, traveling salesman problem solutions, minimum cut/maximum flow algorithms, spectral graph algorithms, and pretrained transformer models.

Part 2 has the following chapters:

- *Chapter 3, Demographic Data*
- *Chapter 4, Transportation Data*
- *Chapter 5, Ecological Data*

3
Demographic Data

In this chapter, we will work with social data to analyze cultural trends spreading across countries and epidemics spreading through the **African Institute for Mathematical Sciences (AIMS)** student population first introduced in *Chapter 2*. We'll see how cultural similarity, geography, and shared social traits create ties between people and places that influence network structure. We'll also introduce a few important structures commonly found in networks. Finally, we'll examine how the spread of ideas and disease is influenced by network structure through our cultural trend- and epidemic-spreading examples.

By the end of this chapter, you'll understand how social factors influence social tie formation, how the spread of ideas and disease is influenced by network structure, and how to spot important features in a plotted network. We'll build on these ideas in later chapters. For now, let's look at populations and their defining characteristics.

Specifically, we will cover the following topics in this chapter:

- Introduction to demography
- Francophone Africa music spread
- AIMS Cameroon student network epidemic model

Technical requirements

You will require Jupyter Notebook to run the practical examples in this chapter.

The code for this chapter is available here: `https://github.com/PacktPublishing/Modern-Graph-Theory-Algorithms-with-Python`

Introduction to demography

Demography, originally formulated as the study of vital statistics about a population, measures individual characteristics across a population of interest. Vital records about a population are important for public policy creation, epidemic tracking, urban planning, and many other government tasks critical for infrastructure and growth. John Snow pioneered the field of demography during the 1854 cholera outbreak in London, plotting geographic data and shared characteristics of infected individuals to determine the source of the cholera outbreak (a water pump). Since then, demography has played a prominent role in medical research, policy planning, education reform, and other applications of the social sciences to society's needs.

In this chapter, we'll introduce some important aspects of demographic research and how we can use shared demographic and geographic factors to create networks by connecting vertices with edges based on shared geographies and demographic factors.

Demographic factors

Demography lends us many potential measurements to study populations and their subgroups in a way that allows us to infer possible social ties between individuals in those populations and subgroups. For instance, we may wish to understand distributions of gender, educational attainment, religion, socioeconomic status, or country of origin in a city such as Kigali, Rwanda, Bangalore, India, or Xi'an, China, over a period of city growth. Often, we plot results in simple charts to show changes over time, as shown in *Figure 3.1*, which depicts the 30-year change in Rwanda's religious composition:

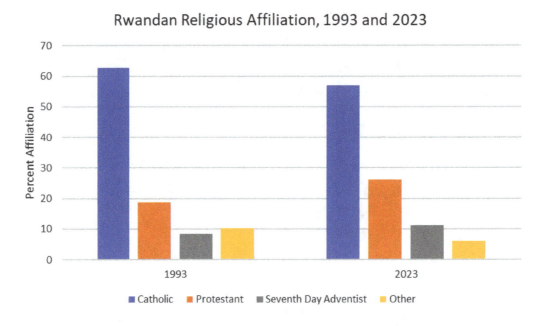

Figure 3.1 – A chart comparing religious affiliation in Rwanda in 1993 versus 2023

Demographic factors vary as populations change, and they also vary across subgroups within populations. While individuals may share demographic factors such as age, gender, and education level, their interests, personality, and upbringing may lead them to different cultural subgroups within a larger population, say within Cape Town, South Africa. Jess may gravitate toward punk rock music and culture. Darla may enjoy socializing and parties. Sadako and Brigette may enjoy nature and artistic pursuits:

Figure 3.2 – Four young women of similar age, socioeconomic status, and educational background who belong to different subgroups in their city

While these young women share some demographic factors that make it likely that they would meet and form social ties in their city, their subgroup memberships make it unlikely that the bartender who enjoys parties would meet and create a friendship with the artistic girl who creates her own clothes to reflect nature. However, if these four young women moved abroad to Dubai, UAE, their shared upbringing in the same city in a foreign country makes it likely that they would meet and form friendships as they navigate a new country together.

In most cases, demographic factors play a role in the formation of social ties, but they do not ensure those ties exist. However, in the absence of surveys about friendships or social media data that we can scrape, demographic factors provide evidence of potential ties that exist among individuals who share common backgrounds. We'll return to this notion later in the chapter when we create a new AIMS Cameroon student network based on demographic factors rather than endorsed interactions among students.

Geographic factors

Geography also plays an important role in determining the formation and likelihood of social ties. Individuals within the same region are more likely to meet as they go out to dinner at the same restaurants, hike in the same areas, and join region-based interest groups. For instance, consider three areas—a suburban setting with bigger houses and lush yards, a densely populated city, and a very rural community with few houses (shown in *Figure 3.3*):

Figure 3.3 – An illustration of three different geographic areas

Perhaps the areas shown in *Figure 3.3* all reside in Belgium, giving them a shared national culture. While we'd expect social ties to form at a higher likelihood among individuals in each geographic area in *Figure 3.3* than across areas, social ties across these regions of Belgium are more likely to form than they are to form with individuals in Nairobi or Tehran, shown in *Figure 3.4*:

Figure 3.4 – An illustration of women in Nairobi and Tehran, respectively

People who share geographies, whether at the country, city, or neighborhood level, are more likely to know each other through regional interactions. However, it is possible that a researcher in Brussels might meet a researcher in Nairobi or Tehran through an academic conference or professional social media platform such as **ResearchGate** or **LinkedIn**. The rise of a global economy and global social media platforms has lessened the impact of geography on social ties, but it is still more likely that those in the same geographic area will form social ties, particularly when they share demographic characteristics.

However, globalism, particularly for those in educated professions, has changed social tie-formation processes, and for those studying social networks among professionals, it is wise to consider other demographic factors associated with the profession rather than just considering geography. The COVID-19 pandemic and the subsequent rise of online/hybrid work, conferences, and networking events widened many professionals' social networks beyond local geographies. Still, those who share a common background geographically tend to share a culture, making those ties more likely even if the geographies of those involved have changed since the pandemic.

Homophily in networks

As we have seen, demographic and geographic factors drive connections between individuals in a population. Since the advent of the knowledge economy, most geographic areas have concentrated on certain demographic factors as people choose to live with others like themselves who work in the same regions where they work, have similar levels of wealth as their levels of wealth, and come from similar educational backgrounds. Similar individuals are more likely to meet, more likely to marry, and more likely to belong to the same groups, as well.

Let's consider three young women (Mariko, Rei, and Ryoko) in Tsuwano, Japan, who share an age group, have similar levels of education, and come from traditional backgrounds, as shown in *Figure 3.5*:

Figure 3.5 – Mariko, Rei, and Ryoko outside Tsuwano, Japan

Based on geography and shared demographic factors, we would expect Mariko, Rei, and Ryoko to form social ties should they meet; given the size of their town, it is likely that they have met already. In network science, this tendency for vertices to form edges with other vertices that share important traits such as demographic or geographic factors is termed **homophily**.

Homophily plays an important role in link prediction within social networks, suggesting potential connections to an individual based on both shared demographic/geographic factors and shared mutual connections. For instance, consider the potential for a connection between Louis and Jean, two data scientists in Paris who are not known to be connected professionally but who share twenty mutual colleagues and many demographic factors: age group, educational background, the arrondissement in which they live and work, and profession. It is likely that they will form a professional connection if they meet at a professional event, and it is possible that they already know each other but have not connected on social media platforms yet. It is much more likely that Louis knows Jean than Patrick, a much older professional in real estate who lives in another arrondissement, never attended graduate school, and never attends tech conferences.

Beyond link prediction, homophily is also useful in clustering networks. Rather than cluster vertices based on network connectivity through network-specific clustering methods (detailed in a later chapter), we can cluster vertices by demographic and geographic factors we've collected with k-means or density-based clustering. While this is more common in social networks, it is possible to apply this clustering to gene networks, sales networks based on geography, or any other type of network.

Let's see homophily in action with two examples, the first highlighting how homophily influences the spread of music trends in Francophone countries in Africa and the second showing how an epidemic might spread through our AIMS student network based on social interactions through students' fields of study and home countries.

Francophone Africa music spread

Countries that share a common language tend to share common music tastes, literature, and other markers of culture, as well as trade together, provide job opportunities across borders, and share economic ties. In Africa, many countries' histories include colonial periods in which they were under either English or French control; many countries still use English or French as a lingua franca for administrative purposes. To understand music spread across Francophone countries, we'll create a network linking countries with more than 10% French-speaking populations that share a border. While other countries, such as Mauritius or Madagascar, have large French-speaking populations, they are geographically isolated from other parts of Francophone Africa. Trends in music or literature may spread from other Francophone countries through online sources or radio stations; however, regional concerts and events will probably not contribute to spreading as readily as they will to countries that share a border.

We first consolidated online information about country borders and estimates of French-speaking populations. Let's create our Francophone country network with Script 3.1 based on the information we found online:

```python
#load needed packages
import numpy as np
import networkx as nx
import matplotlib.pyplot as plt


#create Francophone country network
G = nx.Graph()
G.add_nodes_from([1, 23])
G.add_edges_from([
    (23,13),(13,11),(12,13),(13,18),(13,1),(12,3),(12,18),
    (3,22),(3,18),(3,21),(3,20),(18,1),(18,14),(18,17),(18,21),
    (21,17),(21,14),(17,14),(14,1),(14,6),(14,5),(6,5),(1,8),
    (8,16),(8,7),(16,7),(16,2),(16,4),(16,19),(19,4),(4,2),
    (7,10),(2,10),(10,15),(10,9)])
```

Now that we have the 23 Francophone countries that border at least 1 other Francophone country, let's visualize our country network by adding to Script 3.1:

```python
#plot the Francophone network
import matplotlib.pyplot as plt
G.nodes[1]['country'] = 'Niger'
G.nodes[2]['country'] = 'Republic_of_Congo'
G.nodes[3]['country'] = 'Senegal'
G.nodes[4]['country'] = 'Gabon'
G.nodes[5]['country'] = 'Benin'
G.nodes[6]['country'] = 'Togo'
G.nodes[7]['country'] = 'Central_African_Republic'
G.nodes[8]['country'] = 'Chad'
G.nodes[9]['country'] = 'Rwanda'
G.nodes[10]['country'] = 'Democratic_Republic_of_Congo'
G.nodes[11]['country'] = 'Morocco'
G.nodes[12]['country'] = 'Mauritania'
G.nodes[13]['country'] = 'Algeria'
G.nodes[14]['country'] = 'Burkina_Faso'
G.nodes[15]['country'] = 'Burundi'
G.nodes[16]['country'] = 'Cameroon'
G.nodes[17]['country'] = 'Cote_dIvoire'
G.nodes[18]['country'] = 'Mali'
G.nodes[19]['country'] = 'Equatorial_Guinea'
G.nodes[20]['country'] = 'The_Gambia'
```

```
G.nodes[21]['country'] = 'Guinea'
G.nodes[22]['country'] = 'Guinea_Bissau'
G.nodes[23]['country'] = 'Tunisia'
labels = nx.get_node_attributes(G, 'country')
nx.draw(G, labels=labels, font_weight='bold')
```

Figure 3.6 shows a plot of our Francophone countries and their links through shared borders:

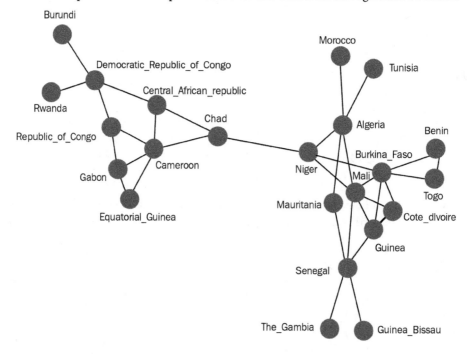

Figure 3.6 – Visualization of the Francophone country network

> **Note**
>
> When working with long labels, the plot of a graph with labels will include overlapping labels or labels that are cut off. However, *Figure 3.6* shows some of the important properties of our network: namely that we have two countries (Chad and Niger) that connect two separate networks (Central African Francophone countries such as Gabon or Burundi and Western/ Northern African Francophone countries such as Senegal or Morocco). As we'll see when we create our model of music trend spread, *bridges* that connect different parts of a network, such as Chad and Niger in the Francophone country network, govern the spreading process from one subnetwork to another.

To model the spread of trends—or diffusion of ideas or spread of disease—we turn to a field of mathematics called differential equations. **Differential equations** relate functions and changes in those functions over time (their derivatives) across a system of interest. For instance, differential equations can model heat flow across a machinery part, opinion changes in a population, gravity's impact on an object in motion, movement of sound through a room, or weather patterns over history (among many others!).

Differential equations can include first-order derivatives, or changes in the slope of a function of interest, second-order derivatives, related to the curvature of the function, and even higher-order derivatives. For spreading processes on networks, we'll limit our discussion to differential equations with first-order derivatives. However, any type of differential equation can be defined and run using the software we'll be using on our NetworkX network.

Sometimes, differential equations include multiple variables in the function of interest; these equations are called **partial differential equations**. Many real-world processes, including epidemic spread models, are modeled with partial differential equations. In addition, many real-world processes have a component of randomness or a dependency on where the process starts on an object or a network. On networks, it's common to run a partial differential equation many times with the same or varying starting points of the process. This allows for a range of possible outcomes.

Now that we know a bit about differential equations, let's dive into our model of interest for spreads of behavior, ideas, and epidemics: the **susceptible-infected-resistant (SIR)** model. The SIR model was originally developed to study epidemics (in 1927), with components related to susceptible individuals in a population (not yet infected but able to be infected), infected individuals (those currently infected with the disease), and recovered or resistant individuals (those who had been infected or were immune and could not be infected).

As individuals mix in the population, infected individuals can pass on their infection to those who are susceptible with a certain probability. For our model, the probability of infecting a susceptible individual is given by the parameter *beta*. In practice, this is either known from estimates in the literature regarding behavior or disease spread probability or is estimated within a range of best guesses, with models run between those best guesses to estimate potential spread within that range of possibilities. Once an individual is infected, that individual has a probability of recovery, given by the parameter *gamma*. Once recovered, our model does allow for reinfection. In a **susceptible-infected-recovered-susceptible (SIRS)** model, however, individuals who have recovered can be reinfected after a period of time. In a **susceptible-infected-susceptible (SIS)** model, recovery is replaced by an immediate chance of reinfection rather than recovery.

One important property of epidemic spread is the so-called **reproductive number** (R_0), or the number of secondary infections caused by an infected individual. We can calculate R_0 by dividing our `beta` parameter value by our `gamma` parameter value. If R_0 is above 1, the epidemic is expected to spread across the population in its initial phases, signaling the potential for a large epidemic. Many infectious disease outbreaks have an R_0 value above 1, including recent Ebola outbreaks (R_0~1.5), the 1960s measles outbreak in Ghana (R~14), and the 1918 Spanish flu (R_0~1.5 initially and ~3.5 in later waves). The R_0 value for COVID-19 varied dramatically between variants and countries, making it difficult to obtain a standard R_0. However, one estimate across Africa puts COVID-19's estimated R_0 at 2.0-9.7 for most countries.

SIR models work a bit differently on networks. An infected vertex typically infects only neighbors in the network, creating a proximity effect. In our network, the bridging properties limit the chance of spreading from one subnetwork to another, as infection rates of a bridge vertex and its adjacent vertices are lower than spread through a hub of mutually connected vertices. In general, bridges are important vertices in networks with respect to information, ideas, or epidemic spread and are often intervention targets. Vaccination campaigns in a disease epidemic to limit spread, targeting in a marketing campaign to increase product adoption rates, and capturing in a criminal communications network are all real-world examples of bridge targeting to impact spread dynamics on a network of interest.

Let's dive a bit deeper into the geometry of networks, including hubs and bridges. **Hubs** are regions of high connectivity, where many vertices connect to many other vertices. Geometrically, densely connected regions such as hubs have high curvature and provide many possible paths for spreading processes or random walks on the network. In our Francophone Africa network, Cameroon, Senegal, and Mali all show hub properties.

Note, however, that Senegal has many edges to other vertices that do not have large numbers of edges associated with them. While Senegal serves as a hub owing to its connectivity, it does not make an ideal spreading target, as the process will die in the fairly isolated vertices surrounding it. Cameroon and Mali, however, are hubs connected to other vertices that serve as either hubs or bridges, allowing the process to continue spreading with high probability. **Centrality metrics** measure how central a vertex is to the network, creating different types of network hub measurements that can find vertices that function as Mali/Cameroon or as Senegal within the network. We will revisit these centrality measurements in later chapters, where we will assess network properties with igraph.

Bridges work a bit differently but also contribute important geometric features to networks. Niger and Chad serve as bridges in our Francophone Africa network, connecting hubs that share no other edges between them. Technically, a type of centrality called **betweenness centrality** measures this bridging property by computing the shortest paths between each vertex pair (representing the most direct communication route) and counting the number of shortest paths that include a given vertex. Those vertices with many of the shortest paths passing through them act as bridges between different parts of the network. Geometrically, bridges serve as crossing points from one part of a network to another with respect to flow. Our music spread from Gabon to West Africa would not be possible without bridges such as Chad and Niger connecting North and West Africa with Central Africa, as the trend must spread through those countries to leave Central Africa.

For our music spread across Francophone Africa, let's consider a SIR model that begins with 5% of our countries listening to a new Rumba artist. We'll say that our music trend has a 25% chance of spreading to a neighboring country (our beta parameter) and a 10% chance of fizzling out as a trend (our gamma parameter). This gives us $R_0 \sim 2.5$, which means our music trend should spread well initially.

Let's download and import the ndlib package by adding to Script 3.1:

> **Note**
>
> Some dependencies may exist for packages that will need to be installed if they are not already on your machine.

```
#install and import epidemic models
#you may need to install bokeh and restart the notebook if it is not
#already installed
!pip install ndlib
import ndlib.models.epidemics as ep
```

Now, we can select the epidemic model we wish to implement and set up its parameters according to our proposed music spread epidemic by adding to Script 3.1:

```
# Model Selection
model = ep.SIRModel(G)

# Model Configuration
import ndlib.models.ModelConfig as mc

config = mc.Configuration()
config.add_model_parameter('beta', 0.25)
config.add_model_parameter('gamma', 0.1)
config.add_model_parameter("fraction_infected", 0.05)
model.set_initial_status(config)
```

Now that we have our model imported and configured according to our proposed spread dynamics, we can simulate this on our Francophone country network. We'll include 50 time periods—perhaps a weekly time period denoting spread over the course of about a year. Let's add to Script 3.1 to simulate the spread:

```
# Simulation
iterations = model.iteration_bunch(50)
trends = model.build_trends(iterations)
```

If your run gives **Bokeh** package code warnings but runs the iterations, the code has run correctly. Some **Bokeh** versions and operating systems will provide warnings about model runs. We can now plot the dynamics of this spread across our network using the bokeh library:

```
#visualize spread dynamics
from bokeh.io import output_notebook, show
from ndlib.viz.bokeh.DiffusionTrend import DiffusionTrend
```

```
viz = DiffusionTrend(model, trends)
p = viz.plot(width=500, height=400)
show(p)
```

This should give you a plot similar to *Figure 3.7*, which shows the fraction of susceptible, infected, and recovered/removed countries in our music trend epidemic. Note that our peak expected number of countries influenced by this new Rumba artist is about a third of our countries, and the influence of this new artist lasts for over half the year. This is a fairly successful outcome for our new artist:

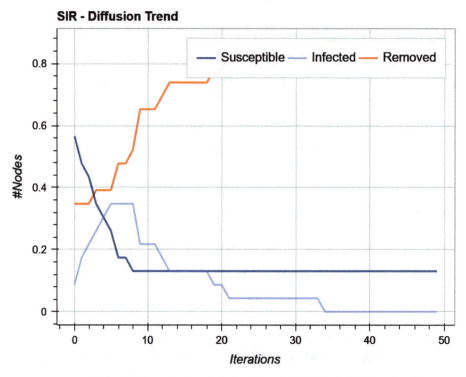

Figure 3.7 – Music trend spread across the Francophone country network

Now, let's modify our gamma parameter to equal 0.3, giving an R_0 of less than 1. We'd expect our trend to fizzle out very quickly with these dynamics. Modifying Script 3.1 under this new scenario yields a much different dynamic (*Figure 3.8*):

```
# Model Configuration
import ndlib.models.ModelConfig as mc

config = mc.Configuration()
```

```
config.add_model_parameter('beta', 0.25)
config.add_model_parameter('gamma', 0.3)
config.add_model_parameter("fraction_infected", 0.05)
model.set_initial_status(config)

# Simulation
iterations = model.iteration_bunch(50)
trends = model.build_trends(iterations)

#visualize spread dynamics
from bokeh.io import output_notebook, show
from ndlib.viz.bokeh.DiffusionTrend import DiffusionTrend

viz = DiffusionTrend(model, trends)
p = viz.plot(width=500, height=400)
show(p)
```

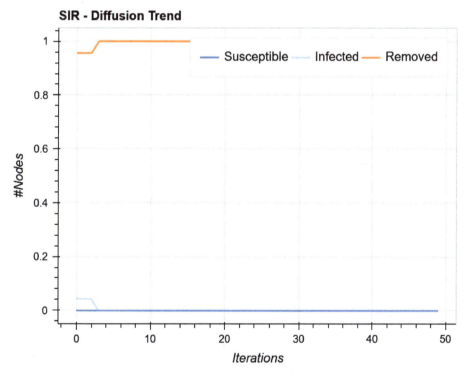

Figure 3.8 – Music spread across the Francophone country network under the new gamma parameter

Figure 3.8 indicates that the new Rumba artist's music does not spread much to neighboring countries and fizzles out soon after the release of the album. It's unlikely that the artist will be known outside their home country and the ones who have heard the album will forget the songs quickly. This is a much different outcome for our artist than the original scenario.

AIMS Cameroon student network epidemic model

To understand how influenza might spread through a social network, let's return to our AIMS Cameroon student network. Recall that we used survey data about interactions to create connections between students. This allowed us to connect students who interacted regularly. However, consider the case where survey data about interactions did not exist. How might we infer possible connections between students?

We have information such as country of origin, field of study, and age of the students. Students from the same countries likely share a common culture—perhaps the same local language, the same favorite comfort foods, or the same favorite musicians.

Similarly, students in the same field of study are likely to spend time together in and out of class, given their shared courseload and career interests. Study groups, group projects, and professional events in their fields likely bring students together on a regular basis. Should an infectious outbreak such as influenza or COVID-19 happen at any of these gatherings, it will likely spread within fields of study.

Let's create a new AIMS Cameroon student network by connecting students who come from the same country or share a field of study. We'll load the dataset with `Script 3.2`:

```
#import needed packages
import pandas as pd
import networkx as nx
import matplotlib.pyplot as plt
import numpy as np
import math

#reading the dataset
fichier = pd.read_csv("C:/users/njfar/OneDrive/Desktop/AIMS_data.csv")
data = pd.DataFrame(fichier)
```

Now, let's set up our network and some preliminary vectors and quantities to help us generate connections between students, including the number of students and their attributes, by adding to `Script 3.2`:

```
#intializing the social network
aimsNetwork = nx.Graph()

#populating the network with nodes and edges
for i in range(len(data["Name"])):
```

```
        aimsNetwork.add_node(
            data["Name"][i], Age=data["Age"][i],
            country=data["Country"][i],Field=data["Field"][i],
            background=data["Background"][i])

#define length
N = len(aimsNetwork.nodes())

# one can build the adjacency matrix
AIMS_by_Country_or_Field = nx.Graph()

#define objects
AIMS = aimsNetwork
students = list(AIMS.nodes())
```

Now, we can connect students according to shared home country or field of study by looping through our attributes to build a graph in NetworkX by adding to Script 3.2:

```
#create edges
for i in range(N-1):
    for j in range(i+1,N):
        sti = students[i]
        stj = students[j]
        if AIMS.nodes[sti]['Field'] == AIMS.nodes[stj]['Field'] or
            AIMS.nodes[sti]['country'] == AIMS.nodes[stj]['country']:
            AIMS_by_Country_or_Field.add_edge(sti,stj)
```

Now that we have our network, let's visualize the connections between students at AIMS:

```
#create plot
Degree=AIMS_by_Country_or_Field.degree()
var = [10*k[1] for k in list(Degree)]
plt.figure(figsize=(20,20))
nx.draw_random(AIMS_by_Country_or_Field, font_size=10,
    node_size = var, with_labels=True, node_color="red")
plt.show()
```

This should give you a plot of a highly connected student network, shown in *Figure 3.9*:

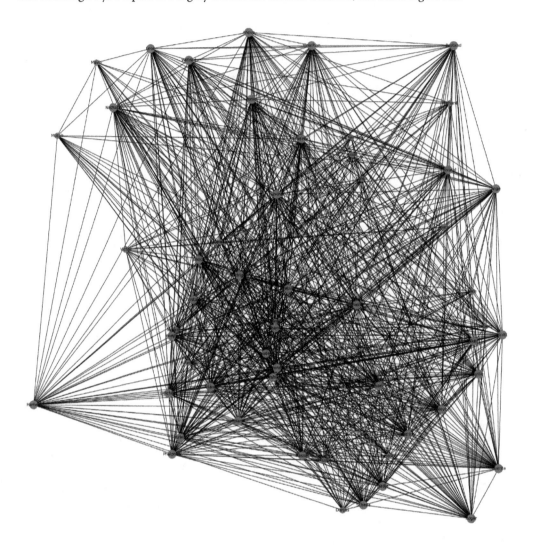

Figure 3.9 – AIMS Cameroon student network connected by country of origin and field of study

Figure 3.9 shows a densely connected network, suggesting many students are likely to interact with many other students. This network does not have many bridges, but most vertices appear to be hubs. In very dense networks, spreading processes tend to move more quickly. The shortest paths between vertices in this network will be relatively short compared to the ones in the Francophone Africa network, as most vertices are connected or nearly connected to each other. The **diameter** of a network, the longest-shortest path, represents the maximum shortest travel route for information or disease to spread between two vertices. A small network diameter (or a small average shortest path length) suggests efficiency in spreading processes, such as information, trends, or disease. Thus, we'd expect our AIMS Cameroon student network to be more vulnerable to spreading processes than our Francophone Africa network. Technically, the diameter provides bounds for differential equations, such as our spreading process, that are defined on the network.

An epidemic with an R_0 over 1 on this network would likely spread quickly and infect most of the network. Let's run an epidemic with the same `beta` and `gamma` parameters as we did on our Francophone network and plot the results to see how an epidemic similar to the Spanish flu might circulate through our student population by adding to `Script 3.2`:

```
#Run the simulated epidemic on the AIMS Cameroon student network
model = ep.SIRModel(G)
config = mc.Configuration()
config.add_model_parameter('beta', 0.25)
config.add_model_parameter('gamma', 0.1)
config.add_model_parameter("fraction_infected", 0.05)
model.set_initial_status(config)
iterations = model.iteration_bunch(50)
trends = model.build_trends(iterations)
viz = DiffusionTrend(model, trends)
p = viz.plot(width=500, height=400)
show(p)
```

Figure 3.10 shows the dynamics of the epidemic, including many infected students at the epidemic peak and nearly the entire network being infected by the time the epidemic abates by week 32:

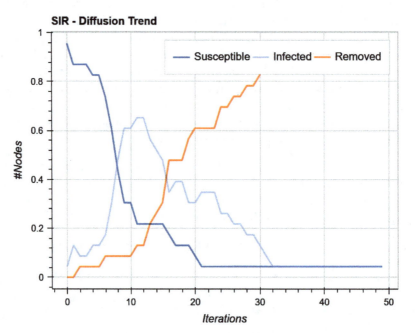

Figure 3.10 – An epidemic with beta = 0.25 and gamma = 0.1 spreading
through the AIMS Cameroon student network

Figure 3.10 shows a very severe epidemic that would probably necessitate the shutdown of classes and isolation of infected students. In *Chapter 8*, we'll consider specific strategies that you can leverage to decrease the impact of epidemics at a population level, including the removal of key vertices that bridge populations or create hubs. For now, let's modify Script 3.2 to run the second type of epidemic, where R_0 is less than 1 (beta = 0.25 and gamma = 0.3):

```
#Run the simulated epidemic on the AIMS Cameroon student network
model = ep.SIRModel(G)
config = mc.Configuration()
config.add_model_parameter('beta', 0.25)
config.add_model_parameter('gamma', 0.3)
config.add_model_parameter("fraction_infected", 0.05)
model.set_initial_status(config)
iterations = model.iteration_bunch(50)
trends = model.build_trends(iterations)
viz = DiffusionTrend(model, trends)
```

```
p = viz.plot(width=500, height=400)
show(p)
```

Figure 3.11 shows a very different type of epidemic that does not start infecting large numbers of students and quickly fizzles out:

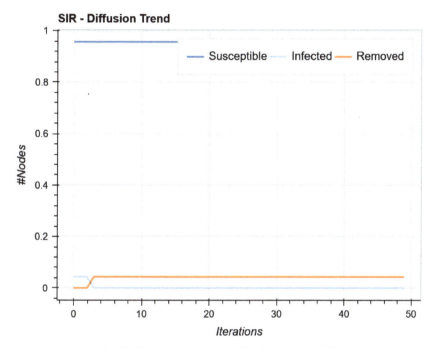

beta: 0.25, gamma: 0.3, fraction infected: 0.05, tp rate: 1

Figure 3.11 – An epidemic with beta = 0.25 and gamma = 0.3 spreading
through the AIMS Cameroon student network

As we can see from *Figure 3.11*, even densely connected networks are not threatened by certain types of epidemics. Without the right conditions to spread effectively—either through network structure or through disease characteristics impacting spread potential—epidemics fizzle out in populations. However, it is possible for a virus with initial dynamics, such as the one shown in *Figure 3.11*, to mutate into something more infectious.

As mentioned earlier in this chapter, the 1918 Spanish influenza strain mutated from an infectious disease to a highly infectious disease, changing R_0 from just over 1 (allowing for spread) to 3.5, heralding a worldwide disaster. In practice, it's important to consider a variety of realistic parameter values when simulating epidemics circulating through networks of interest, as a small difference in parameters can lead to a very different outcome; epidemic planning works best when a wide variety of scenarios are simulated.

Summary

In this chapter, we introduced demographic factors and how they influence network development through homophily. We also introduced the concepts of hubs and bridges. We then turned to SIR models to show trends and disease spread through social networks based on spreading factors in the differential equation models. We considered two networks: a geographic network of neighboring Francophone countries and the AIMS Cameroon student network with inferred edges based on demographic factors. We then examined the role of R_0 on both networks through simulations of epidemics by changing epidemic parameters; this demonstrated that specific conditions are needed for trends or diseases to spread across a network. We'll revisit these ideas and strategies to influence spreading behavior again in *Chapter 8*, when we dive into centrality measures in more detail with respect to dynamic network metrics change over time as individuals change their connections with others in the network over time.

In the next chapter, we'll consider the role of networks in transportation logistics and how to represent transportation route data as a network.

References

Aiello, L. M., Barrat, A., Schifanella, R., Cattuto, C., Markines, B., & Menczer, F. (2012). Friendship prediction and homophily in social media. *ACM Transactions on the Web (TWEB), 6(2), 1-33.*

Cooper, I., Mondal, A., & Antonopoulos, C. G. (2020). A SIR model assumption for the spread of COVID-19 in different communities. *Chaos, Solitons & Fractals, 139, 110057.*

Das, K., Samanta, S., & Pal, M. (2018). Study on centrality measures in social networks: a survey. *Social network analysis and mining, 8, 1-11.*

Estrada, E., Kalala-Mutombo, F., & Valverde-Colmeiro, A. (2011). Epidemic spreading in networks with nonrandom long-range interactions. *Physical Review E, 84(3), 036110.*

Iyaniwura, S. A., Rabiu, M., David, J. F., & Kong, J. D. (2022). The basic reproduction number of COVID-19 across Africa. *Plos one, 17(2), e0264455.*

McPherson, M., Smith-Lovin, L., & Cook, J. M. (2001). Birds of a feather: Homophily in social networks. *Annual review of sociology, 27(1), 415-444.*

Rodrigues, F. A. (2019). Network centrality: an introduction. *A mathematical modeling approach from nonlinear dynamics to complex systems, 177-196.*

Smith, H. L. (2003). Some thoughts on causation as it relates to demography and population studies. *Population and Development Review, 29(3), 459-469.*

Weiss, H. H. (2013). The SIR model and the foundations of public health. *Materials mathematics, 0001-17.*

4
Transportation Data

This chapter tackles transportation logistics, which involves the movement of supplies or goods from one location to another. We'll introduce a goods delivery problem to find the optimal routing of supplies to minimize the delivery time and cost to deliver the goods. We'll explore shortest paths, optimal routes to visit all necessary locations, and scaling algorithms to large networks. Further, we'll examine caveats to simple distance weightings to calculate route optimality, considering delivery hazards on routes that can influence optimality.

When you have finished this chapter, you'll understand how to frame transportation problems as network problems and scale them to very large routing networks using Python.

Specifically, we will cover the following topics in this chapter:

- Introduction to transportation problems
- Shortest path applications
- Traveling salesman problem
- **Maximum flow/minimum cut (max-flow min-cut)** algorithm

Let's get started with some basic problems in transportation logistics.

Technical requirements

You will require Jupyter Notebook to run the practical examples in this chapter.

The code for this chapter is available here: `https://github.com/PacktPublishing/Modern-Graph-Theory-Algorithms-with-Python`

Introduction to transportation problems

Physical goods and supplies are important in many industries, and the movement of goods between locations represents an important problem in industries such as consumer-packaged goods, retail, military, and manufacturing. **Supply chain logistics**—the science of acquiring, transporting, and storing resources—influences many aspects of business in these industry sectors. Without goods to sell, a company cannot turn a profit. Without materials to manufacture goods, products cannot be made or transported to vendors.

During the COVID-19 pandemic, many critical supply chain routes shut down, leaving long waits for goods in many parts of the world or facing higher prices for necessities such as food. Crises such as the Ukraine war can leave entire countries short of food, creating humanitarian crises in other areas. Many supply chain logistics problems can be formulated through the lens of network science, and graph theory offers several useful tools to plan out the best routes to stock goods or materials across locations. In this chapter, we'll learn more about supply chain logistics and leverage tools from graph theory to plan supply routes between grocery stores.

Paths between stores

A common supply chain logistics problem retailers face is the transportation of goods across several stores in an area. For instance, consider a suburban area with five grocery stores belonging to the same chain spread out across the area, as depicted in *Figure 4.1*:

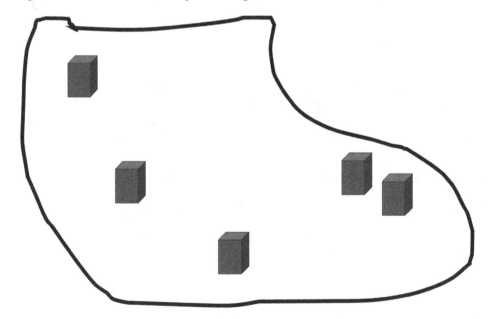

Figure 4.1 – A metro area with five grocery stores from the same chain

The metro area of *Figure 4.1* shows stores that are nearby (right corner) and some that are further away from each other (such as the ones on the left). Perhaps we are transporting fresh fruit and vegetables from a nearby farm to each of the five stores. Without considering streets, fuel stops, or other considerations (we'll discuss these later in the chapter), it seems like a route between stores would be easy to define. One such path is shown in *Figure 4.2*:

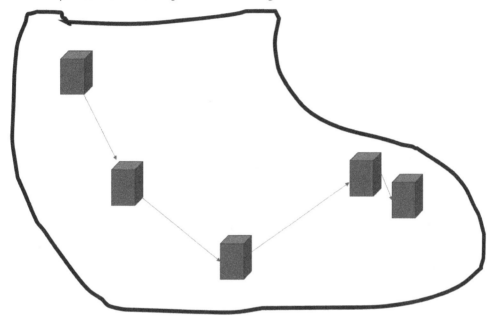

Figure 4.2 – A routing plan to deliver fruit and vegetables to each of
the grocery stores in the hypothetical suburban area

Contrary to *Figure 4.2*, in a real situation, there are likely to be many routes between stores a truck could take. Alternative streets and combinations of routes might exist. The truck may need fuel between stores, and the locations of gas stations might influence which route is ideal. Roads may be closed for construction. Afternoon thunderstorms might flood some of the streets on good routes, creating hazards that the truck needs to avoid. Some stores might be out of oranges and grapefruit, giving them priority for early deliveries so that customers can purchase these fruits.

Over larger distances, the question of transportation mode also arises. Shipping by sea or routing goods by plane might be preferable if obstacles such as oceans or mountain ranges exist between stores that need stocking. Since the world globalized, it is common for goods to travel across continents and oceans (shipping is shown in *Figure 4.3*):

Figure 4.3 – An illustration of a cargo ship leaving port

Many products or parts of products used in the United States and Europe come from China, Japan, or India. Travel by sea can provide cheaper and quicker alternatives to land or air delivery. One of the key problems in supply chain logistics is balancing the needs for cost-effectiveness, **time-to-delivery** (TTD), and the need for bulk goods movements. Let's dive into some of these considerations in more detail.

Fuel costs

One of the pressing problems in logistics today is the optimization of a route with respect to the cost of delivery. Customers don't want to pay high prices for delivery or wait long periods of time for goods to arrive. Since the COVID-19 shutdowns and the Ukraine war, fuel prices have increased, and some routes for goods are no longer available. This drives up costs for manufacturers, businesses that need to move goods to physical locations, and consumers buying goods from businesses.

Ground routes, particularly when coupled with electric delivery vehicles (shown in *Figure 4.4(a)*), can provide a good option for companies operating locally. However, many supply chains depend on parts or goods located overseas, and options such as shipping (shown in *Figure 4.4(b)*) or air travel are necessary:

Figure 4.4 – An illustration of: (a) A semi-truck used to haul goods
domestically and (b) a shipping option to move goods overseas

However, the cost of fuel is rarely the only consideration, as time is critical for any goods that might spoil and as safety concerns related to cargo loss often take precedence over fuel costs in real-world problems. Let's consider some scenarios where TTD is an important consideration in the optimization algorithms used to schedule delivery times, routes, and best modes of transportation.

Time to deliver goods

The first consideration for delivery time involves the urgency factor. Goods may spoil within a certain time frame (such as produce or vaccines), and this often outweighs the cost, as the goods will not be useable outside of the necessary time-to-use period. Sales windows also merit consideration, as Christmas or Ramadan items will not be in as great a demand after the holiday passes. This results in a loss of profits for the business.

Let's consider two tourist cafés, one located in the city of Paramaribo, Suriname, and one located in rural Suriname near a nature adventure camp. The city location (shown in *Figure 4.5 (a)*) has access to a refrigerator to store fruit used in its smoothies; the rural location (shown in *Figure 4.5 (b)*) does not have a refrigerator. Delivery priority on a shipment of fresh fruit from a local farm might prioritize the rural location, as the fruit will spoil faster there:

Figure 4.5 – An illustration of: (a) A café in Paramaribo, Suriname
and (b) a café in rural Suriname near a tourist camp

Besides the problem of spoiled goods and missed promotional periods, timing factors can include barriers to delivery, such as construction or tollways on a route. In addition, time spent in customs can vary dramatically from country to country, and it may be advantageous to take a longer driving route to avoid traveling through a particular country.

Often, it is necessary to mathematically weight routes not only by physical distance but by travel time, including factors such as customs or likelihood of construction.

Navigational hazards

Navigational concerns not only include man-made impediments to delivery; they also include natural ones. Hurricanes can delay shipments through busy ports, such as the Port of Miami. Avalanches in the Alps are a navigational consideration for companies opting for train transport. The shipping industry averages a loss of 10 ships per year to high seas and rogue waves. Some shipping routes, such as those off the southeast coast of South Africa, are known to produce high waves (typically as currents mix with a variety of ocean swells coming from different directions).

Hazard-weighting of paths can be wise when working with optimal travel routes and determining which mode of transportation is most cost-effective and convenient. Most areas of the world contain open source hazard data and material to determine which hazards are present and how likely one is to encounter them. For instance, the World Bank provides the **Climate Change Knowledge Portal (CCKP)** with natural disaster risks during different periods, including downloadable data.

Now that we know some routing considerations, let's dive into a simple example, including five stores in a localized area of Miami needing produce delivered that are all accessible via truck and defined to exclude current construction zones in a time of year when the area does not experience flooding or hurricanes. First, we'll need to understand how we can calculate the shortest paths between vertices in a network, which has deep roots in graph theory.

Shortest path applications

The shortest paths between places and sets of places have a long history in graph theory. Originally, this problem arose from a question about traversing the seven bridges of Königsberg, Germany. In 1736, Leonhard Euler posited that a route that crossed each bridge to a region next to one side of a bridge exactly once did not exist. Indeed, this is the case. If there is one more region than the number of bridges for an odd number of bridges, a trip is possible without traversing bridges more than once. Note that the proof of this is beyond the scope of this book; if you are interested, you can find many proofs online if you search for proofs of the Königsberg bridge problem.

However, problems such as this come up often in the transportation industry and **global positioning system (GPS)** routing solutions and algorithms that calculate shortest paths with or without specific constraints such as the Königsberg bridge problem are common in routing problems today. We often want to visit multiple locations while traversing the fewest roads, bridges, or obstacles possible, and we can formulate this problem much in the way Euler formulated his bridge problem. However, we typically have many different possible routes and locations, making proof much more difficult given alternative routes that exist.

Dijkstra's algorithm calculates the shortest path between vertices on a network. As we've seen before, networks can be unweighted (with 0 or 1 representing the existence of an edge between vertices in the adjacency matrix) or weighted (where edges that exist have numbers not limited to 1 designating their existence and some property between them). In the context of shortest path calculation in geographical data, these weights usually represent distances between locations.

Many paths can exist between vertices. For instance, consider a network with seven vertices. Between vertices one and six, many possible paths exist, including the one shown in *Figure 4.6*. However, the shortest path will always exist, minimizing the distance (physical or other, depending on how edge weights are assigned from a dataset) between any two vertices in the network. When a network is not connected, this may be undefined or set to an infinite distance:

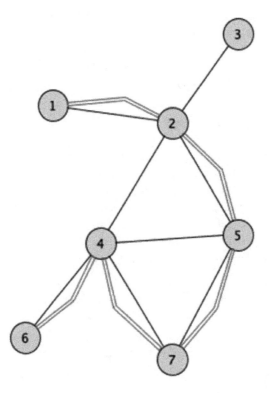

Figure 4.6 – A network with seven vertices and a path from the first vertex to the sixth vertex

Sometimes, the shortest path between vertices (say, between one and seven in our *Figure 4.6* network) may not be unique. Multiple paths of the same length may exist, such as the two shortest paths between vertex one and vertex seven (shown in *Figure 4.7*):

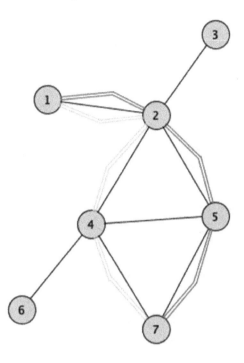

Figure 4.7 – A network with multiple shortest paths between a
pair of vertices (vertex one and vertex seven)

In an undirected network, all edges between vertices can be considered when finding the shortest paths between vertices. As mentioned earlier, if the network is not connected, infinite path lengths may exist. For undirected and unweighted networks, the shortest paths will traverse the fewest edges (again, with infinite values possible if there are no edge paths to connect two vertices). *Figure 4.8* shows a network that is not connected:

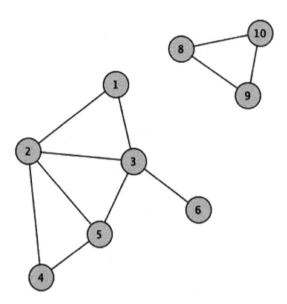

Figure 4.8 – A disconnected network

The shortest path between vertices eight and nine is simply the edge connecting vertex eight to vertex nine. Assuming that this network is not weighted, the shortest distance from this path is one. The shortest path between vertices one and six is given by the edges connecting vertex six to vertex three and vertex three to vertex one. The shortest distance between vertex six and three is, thus, two. However, the shortest path between vertex eight and vertex six does not exist and would be designated as an infinite distance.

Dijkstra's algorithm begins at one of the paired vertices and explores the vertices to which that vertex is connected. The shortest path between that initial vertex and its neighboring vertices is then recorded, minus the initial vertex (as it is already in our path set). The next set of vertices connected to the shortest-distance neighboring vertex is then explored to find the next shortest path that exists. This iterative exploration continues until the other paired vertex is found.

Let's consider how Dijkstra's algorithm can help us find the shortest route between stores. Consider a set of five stores from the same grocery chain in one suburb that may need restocking by the parent chain's supplier. *Table 4.1* gives a summary of the distances between these five stores in miles:

	Store 1	Store 2	Store 3	Store 4	Store 5
Store 1	0	2	2.4	3	3
Store 2	2	0	3.7	1.4	4.3
Store 3	2.4	3.7	0	4.9	0.9
Store 4	3	1.4	4.9	0	5.4
Store 5	3	4.3	0.9	5.4	0

Table 4.1 – A table of mile distances between a chain of stores in a suburb

We can create this network of stores in **NetworkX** through Script 4.1:

```
#import packages
import networkx as nx

#create the Stores network
STRS = nx.Graph()
STRS.add_nodes_from(["store1","store2","store3","store4","store5"])

#define weighted ebunches, which represent lists of edges
e1 = [("store1","store2",2),("store1","store3",2.4),
    ("store1","store4",3),("store1","store5",3)]
e2 = [("store2","store3",3.7),("store2","store4",1.4),
    ("store2","store5",4.3)]
e3 = [("store3","store4",4.9),("store3","store5",0.9),
    ("store4","store5",5.4)]

#add edges
STRS.add_weighted_edges_from(e1)
STRS.add_weighted_edges_from(e2)
STRS.add_weighted_edges_from(e3)

#plot
weight_labels=nx.get_edge_attributes(STRS,'weight')
pos = nx.spring_layout(STRS)
nx.draw(STRS,pos,with_labels=True,)
nx.draw_networkx_edge_labels(STRS, pos,edge_labels=weight_labels)
```

Script 4.1 gives us a network of our five stores with a nice plot summary showing distances between stores, as shown in *Figure 4.9*:

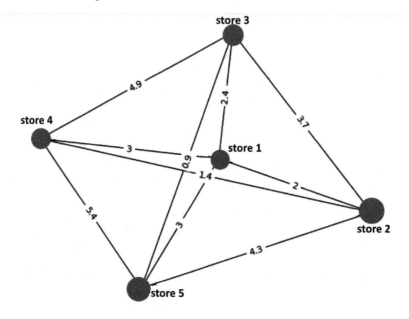

Figure 4.9 – A plot of our five stores and the distances between each of them

We can now create an adjacency matrix based on our network and find the shortest paths between locations by adding to Script 4.1. Let's calculate the shortest distances between **store 1** and each other store in our chain's locations:

```
#create adjacency matrix
adj1 = nx.adjacency_matrix(STRS)
adj1.todense()


#find all shortest paths starting at Store 1
length1, path1 = nx.single_source_dijkstra(STRS,"store1")
length1
```

From this calculation, we derive both the shortest distance between **store 1** and all other stores (length1) and the shortest routes in general (path1). We can see the shortest distances between **store 1** and the other stores from length1. The shortest route to **store 2** takes two miles, while the shortest route to **store 4** or **store 5** takes three miles. **store 3** is 2.4 miles from **store 1**.

One of the major drawbacks of using Dijkstra's algorithm is that it is a blind search algorithm that needs to wander through all possible options to find the shortest paths between each possible vertex pair along the path. That takes a long time for large networks, and in practice, a modification of Dijkstra's algorithm is needed to reduce search time.

The **A* algorithm** (pronounced A-star) modifies Dijkstra's algorithm by using a heuristic function to guide the search (as opposed to visiting each vertex as in Dijkstra's algorithm). Heuristic functions are common in search tasks, as they expedite the process and can modify the search of an algorithm that has failed. The heuristic used in the A* algorithm combines the cost of the path from the starting vertex with an estimate of remaining costs to prioritize vertices closest to the final vertex. One of the main uses for the A* algorithm is to give directions to a user of GPS to find optimal routes to a destination.

In general, the A* algorithm efficiently finds shortest path candidates, particularly in large or dense networks. However, its performance depends on the heuristic used to estimate the cost of reaching a given destination by traversing the network. While Dijkstra's algorithm is guaranteed to find the shortest path, the A* algorithm is not guaranteed to find the shortest path. On large, dense networks, however, it is not feasible to run Dijkstra's algorithm, so A* must be used.

Table 4.2 summarizes the differences between Dijkstra's algorithm and the A* algorithm to guide you on how and when to use each algorithm for shortest path computation:

Shortest path algorithm	Type of search	Result guarantees	Computational speed
Dijkstra's algorithm	Blind search of all vertices	Always finds the shortest path	Slow enough to limit use on dense or large networks
A* algorithm	Heuristic function guiding search	No guarantees to finding the shortest path	Fast enough for most problems

Table 4.2 – Difference between Dijkstra's algorithm and the A* algorithm

Let's consider a different store network, one in which routes do not exist from each store to every other store by creating a secondary store network in NetworkX through `Script 4.2`:

```
#create a different Stores network
STRS2 = nx.Graph()
STRS2.add_nodes_from(["store1","store2","store3","store4","store5"])

#define weighted ebunch
e12 = [("store1","store2",2),("store1","store4",3)]
e22 = [("store2","store3",3.7),("store2","store4",1.4)]
```

```
e32 = [("store3","store4",4.9),("store4","store5",5.4)]

#add edges
STRS2.add_weighted_edges_from(e12)
STRS2.add_weighted_edges_from(e22)
STRS2.add_weighted_edges_from(e32)

#plot
weight_labels=nx.get_edge_attributes(STRS2,'weight')
pos = nx.spring_layout(STRS2)
nx.draw(STRS2,pos,with_labels=True,)
nx.draw_networkx_edge_labels(STRS2,pos,edge_labels=weight_labels)
```

Figure 4.10 shows this new store network, with some stores directly connected to each other and others requiring a route past other stores. Note that **store 5** is the furthest store from the other stores, connected to **store 4** but no other stores:

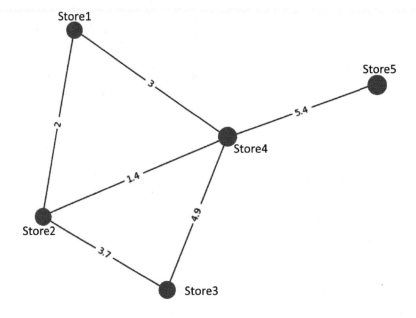

Figure 4.10 – A store network where not every store is connected to another store with a passable route

Let's consider a route from **store 2** to **store 5**. Several possible routes exist (**store 2** to **store 3** to **store 4** to **store 5**, **store 2** to **store 1** to **store 4** to **store 5**, and **store 2** to **store 4** to **store 5**). The shortest possible route may not have the fewest stops along the way if all distances are relatively short. In this case, we can simply look at our network and see that **store 2** to **store 4** to **store 5** is the shortest route. However, in very large networks, this would take a lot of time and likely involve much longer paths.

We can use the A* algorithm in NetworkX to find our shortest route by adding to `Script 4.2`:

```
#find shortest path from Store 2 to Store 5
nx.astar_path(STRS2,"store2","store5", weight='weight')
```

As expected, the A* algorithm finds the shortest path from **store 2** to **store 4** to **store 5**. Again, with a much larger network, we'd need an algorithm such as the A* algorithm to find the shortest route quickly. Even a network of 20 stores with several connections among stores would be problematic to compute by hand from a visualization of the network. We've seen how algorithms can help us find the shortest paths between two specific destinations. Let's now take a look at the shortest paths that include stops at several different vertices. This type of solution is critical to many optimal routing problems and builds on our problem of finding the shortest paths between two different destinations.

Traveling salesman problem

A natural extension of shortest paths is the shortest possible route that stops at each location. For instance, consider a produce truck that needs to stock all five of our stores. The shortest route that will stop at each of our five stores saves time and fuel for the driver and allows produce to arrive at each store in the shortest time frame.

The **traveling salesman problem** seeks to find the shortest route that stops at each location or the shortest route that stops at an arbitrary number of possible locations. In graph theory, this problem (and Euler's problem) is related to cycles of a graph, which define a non-empty path that starts and ends at the same vertex.

In practice, algorithms are needed to find the shortest path, and for large problems, computational time and convergence to a solution can restrict the usage of most algorithms. NetworkX provides the **Christofides algorithm** as a solver, which finds the shortest spanning tree (network structured like a tree with no cycles) and then matches the vertices of the tree to find minimum distances.

Let's find a solution to the traveling salesman problem for our original five-store network with `Script 4.3`:

```
#define traveling salesman algorithm and apply it to the stores's
#networks
tsp = nx.approximation.traveling_salesman_problem
tsp(STRS)
```

This path starts at **store 1**, proceeds to **store 3** (2.4 miles), then **store 5** (0.9 miles), then **store 4** (5.4 miles), then **store 2** (1.4 miles), and finally back to **store 1** (2 miles).

Let's see how this works on our secondary store network by adding to `Script 4.3`:

```
#apply TSP algorithm to the second stores's networks
tsp(STRS2)
```

In this case, we need to visit some stores more than once to visit each store. The solution our algorithm gives us starts at **store 1**, then **store 4** (3 miles), then **store 5** (5.4 miles), then **store 4** again (5.4 miles), then **store 2** (1.4 miles), then **store 3** (3.7 miles), then back to **store 2** (3.7 miles), and finally back to **store 1** (2 miles). This is a much longer route than we need in our original store network (12.1 miles versus 22.6 miles), owing to non-direct paths between many of the stores.

In this section, we reviewed a solution to the traveling salesman problem, where we have a set of vertices that we need to visit and want to find the most efficient way to visit all of them. In the next section, we'll switch from examining shortest path problems and looking at partitioning vertices to maximize travel between vertices in each set (such as maximizing rush-hour traffic flow while shutting down a few routes for maintenance).

Max-flow min-cut algorithm

Aside from shortest paths and routes, transportation logistics sometimes involve city planning to plan, say, roadwork with the least interruption to traffic patterns or supply chains. The goal is to maximize traffic flow through points of interest (say, major intersections or buildings with high volumes of visitors/workers each day) while minimizing which routes are cut off.

In graph theory, the max-flow min-cut algorithm seeks to partition a network to maximize the flow of information through a social network, the flow of traffic in a transportation network, or the flow of material through an electrical or water pipeline network, among others. Typically, there's a starting vertex and an ending vertex with respect to flow, though it is possible to run the algorithm through all possible combinations and aggregate results to maximize flow for the entire network.

Let's consider the example of traffic flow from a dense residential area outside of a city to the downtown area, where most people work. We'd like to plan out work on replacing stoplights at key intersections throughout the city, which are divided by canals. Some roads in the city have more lanes than others, allowing for a higher volume of traffic to utilize those routes. We'd like to minimize disruption to commuters during the workday.

We can create a network of this situation, with the residential area designated as vertex `"a"` and the downtown area as vertex `"g"`. We'll designate the level of traffic the route can handle with the capacity parameter. Let's build an example network in NetworkX with `Script 4.4`:

```
#build the city intersection network
G1=nx.DiGraph()
```

```
G1.add_edge("a", "b", capacity=2.0)
G1.add_edge("a", "d", capacity=0.5)
G1.add_edge("b", "d", capacity=3.0)
G1.add_edge("c", "d", capacity=2.5)
G1.add_edge("c", "e", capacity=1.0)
G1.add_edge("e", "f", capacity=4.0)
G1.add_edge("f", "g", capacity=1.0)


#plot network
capacity_labels=nx.get_edge_attributes(G1,'capacity')
pos = nx.spring_layout(G1)
nx.draw(G1,pos,with_labels=True,)
nx.draw_networkx_edge_labels(G1,pos,edge_labels=capacity_labels)
```

This script should produce a network that looks like the one in *Figure 4.11*:

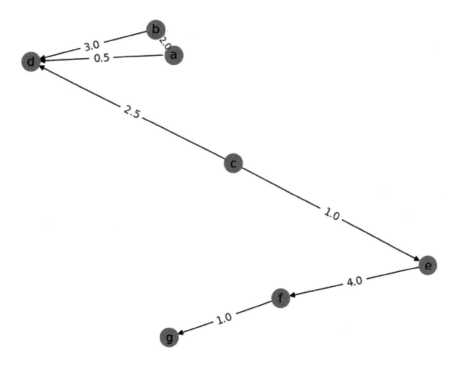

Figure 4.11 – A plot of the traffic network

Now that we have our network, let's apply our max-flow min-cut algorithm and find where we can replace our first traffic light by adding to `Script 4.4`:

```
#find cut-points
partition = nx.minimum_cut(G1, "a", "g")
print(partition)
```

Our partition suggests that our best option cuts the connection between intersections `"c"` and `"d"`. Some commuters may be disadvantaged, but this preserves movement for the largest number of commuters overall.

Summary

In this chapter, we explored transportation problems and routing problems, which come up often in real-world industries. We found the shortest paths between stores with the Dijkstra and A* algorithms on two example store networks. We then considered the traveling salesman problem for optimal route planning on our two store networks to see how connectivity impacts route length. Finally, we considered optimal cuts to maximize flow on a small city network. In the next chapter, we'll tackle clustering on networks by examining ecological data.

References

Chen, L., Kyng, R., Liu, Y. P., Peng, R., Gutenberg, M. P., & Sachdeva, S. (2022, October). Maximum flow and minimum-cost flow in almost linear time. *In 2022 IEEE 63rd Annual Symposium on Foundations of Computer Science (FOCS) (pp. 612-623). IEEE.*

Johnson, D. B. (1973). A note on Dijkstra's shortest path algorithm. *Journal of the ACM (JACM), 20(3), 385-388.*

Kang, N. K., Son, H. J., & Lee, S. H. (2018). Modified A-star algorithm for modular plant land transportation. *Journal of Mechanical Science and Technology, 32, 5563-5571.*

Little, J. D., Murty, K. G., Sweeney, D. W., & Karel, C. (1963). An algorithm for the traveling salesman problem. *Operations research, 11(6), 972-989.*

Liu, S., Münch, F., & Peyerimhoff, N. (2018). Bakry–Émery curvature and diameter bounds on graphs. *Calculus of Variations and Partial Differential Equations, 57, 1-9.*

Rondinelli, D., & Berry, M. (2000). Multimodal transportation, logistics, and the environment: managing interactions in a global economy. *European Management Journal, 18(4), 398-410.*

Rosenthal, W., & Lehner, S. (2008). *Rogue waves: Results of the MaxWave project.*

5

Ecological Data

In this chapter, we'll introduce some methods commonly used in ecological data collection and then explore graph-theoretic clustering methods that can parse out animal populations across monitored geographic areas. This is a very common problem within ecological research, as populations of animals migrate and overlap with other populations of animals. In addition, parsing out healthy and unhealthy ecosystems is a common survey task for conservation and urban planning. We'll analyze two types of ecological data with **spectral clustering** to find groups within snake capture count and ecological survey text data.

By the end of this chapter, you'll understand how to apply spectral clustering to find groups on networks. You'll also see how different implementations of spectral clustering can be formulated, with some algorithms scaling better to large networks than other algorithms.

Specifically, we will cover the following topics:

- Introduction to ecological data
- Spectral graph tools

Let's get started with some background on ecological problems.

Technical requirements

You will require Jupyter Notebook to run the practical examples in this chapter.

The code for this chapter is available here: `https://github.com/PacktPublishing/Modern-Graph-Theory-Algorithms-with-Python`

Introduction to ecological data

Environmental research plays a large role in conservation strategy, climate change monitoring, and farming practices. Data collected might include data on plant type distributions and densities, animal migration patterns, or extent of disease. For instance, farmers may wish to track crop disease across large farms to monitor potential threats to their annual yield from various crops. Conservationists may wish to track endangered animal populations as they migrate through a national park to deploy anti-poaching resources to areas with large native or migratory populations. Let's first explore some data collection methods.

Exploring methods to track animal populations across geographies

Animal populations are frequently mobile, with seasonal changes, weather events, and human interactions driving the movement of animals from one geographic location to another. One of the largest mass migrations of animals occurs in the grasslands of East Africa, where millions of herd animals (such as wildebeests and zebras) trek from the Serengeti to Masai Mara back to the Serengeti each year as rains come and go (*Figure 5.1*). Poaching not only harms animal populations but often fuels other illegal activities, creating problems locally and internationally; understanding migration patterns on the borders of protected areas can position resources to stem poaching:

Figure 5.1 – An illustration of a zebra herd as the rainy season begins

There are many ways to monitor a population's movement across geographies. One way involves tagging animals with satellite tracking tags that upload location data for each tagged animal. This allows researchers to construct paths for each animal and aggregate results across tagged populations (which might include only one species or several species of interest). For instance, in the zebra herd, perhaps a conservation team has tagged 10 of the zebras prior to the rainy season with tags that will track the zebras' location each hour for a period of several weeks. While we would expect all 10 animals to follow the same general path from one park to the next, individuals may migrate at different paces, take slightly different routes, or stop for food and water in different locations. This data allows researchers to identify subpopulations within the herd whose behavior deviates from the herd's majority.

Another common method to track animal populations across geographies is to use trap cameras hidden in locations that are likely to be frequented by species of interest (such as watering holes, clearings in forests, or healthy coral reefs). For instance, suppose public health officials have noticed an increase in king cobra bites in villages near the Atrai River in Bangladesh (*Figure 5.2*) and would like to understand the local king cobra population to enact policies to prevent humans and cobras from interacting.

Perhaps work hours can be adjusted to keep agricultural workers from areas frequented by king cobras at certain times of day or seasons. Officials can set the trap cameras in areas where bites have become more common to capture images of cobras throughout the lifetime of the film used in the trap camera. Several cameras are set throughout the area of interest according to bite densities, and species of venomous snakes, including king cobras, are tracked by time of day, species, and number of individual snakes identified by each camera. Data then informs policies regarding population movement and snake control strategies to mitigate the threat to public safety.

Figure 5.2 – An illustration of king cobra in rural Bangladesh

Exploring methods to capture plant distributions and diseases

Data capture on plant distributions and diseases involves a different sort of approach, as plants generally are not moving across geographies. Manual data collection requires human discernment of plant species and diseases, which may be unreliable and will take long periods of time to capture meaningful data. However, autonomous vehicles and drones offer a way to capture images and videos of plants across wide geographic areas, which can be processed with **convolutional neural networks** (**CNNs**) trained on plant image data of species native to the study area. For instance, Colombian farmers may want to identify instances of coffee rust early to avoid crop loss by running a coffee rust classifier CNN on drone or rover data captured weekly across their regional farms (example shown in *Figure 5.3*). Images showing potential coffee rust could trigger a system alert that provides a geographic location on the farm or plant identification number to farmers.

Figure 5.3 – An illustration of a coffee plant image to scan for coffee rust on a Colombian farm

However, the manual documentation of plants can be a useful tool in studies of areas with potentially new plant species to identify through fieldwork. Much of the Congo and Amazon basins are relatively unexplored, and identifying potential new plant species may yield insights into an important part of the ecosystem, possible medications to treat human diseases, or the identification of threats from an invasive species.

Invasive plants may threaten an ecosystem. New species may herald discoveries of new chemical compounds. Let's consider a botanist doing fieldwork in the Amazon who comes across a new type of mushroom that she hasn't seen before (*Figure 5.4*). To confirm that this is a new mushroom species, our botanist would take photographs, collect a sample for genetic analysis, and log the ecosystem in which she encountered the mushroom.

Figure 5.4 – An illustration of a potentially new species of mushroom

All of these data collection methods create spatial data, where some areas may or may not be physically located next to each other. Species found in groups of geographic areas that are far from each other likely correspond to distinct populations that will be impacted by different human and environmental factors. Treating this data as spatial data and applying a few tools from graph theory can help parse out subpopulations of interest within ecological data analytics.

In many cases, collecting biological samples of plants or placing tracking tags on animals is not feasible. However, drones and other autonomous vehicles can snap pictures of flora and fauna in a region (*Figure 5.5*), allowing researchers to identify species from the images through a special type of deep learning algorithm and build a network from the deep learning results.

Figure 5.5 – An illustration of a forest stream, from which flora can
be identified through deep-learning algorithms

Now that we know how to collect ecological data, let's explore a few tools from graph theory that will help us analyze ecological data that has been collected across different geographical regions.

Spectral graph tools

The **adjacency matrix** and **degree of vertices** in the adjacency matrix contain information about the connectivity of vertices within the network. For an undirected network, the **Laplacian matrix** of the network is found by subtracting the adjacency matrix from the degree matrix (i.e., $L = D - A$, where D is the degree matrix and A is the adjacency matrix). This matrix contains information about many important properties of the network, including the number of connected components, the sparsest cut that can separate the network into separate components, and the strength of connection within the network overall.

Connectivity is an important concept in graph theory. A **connected network** is one in which there is a path between all pairs of vertices; a **disconnected network** is one in which some vertices do not have paths to all other vertices. Connected networks can be separated by cutting edges; the **minimum cut set** is the set of the fewest edges that need to be removed to separate a network into two pieces to disconnect a connected network. Not only do these values connect to theoretical results in graph theory related to spreading processes and other differential equations defined on the network, but this information also tells us about how to separate groups on the network.

To obtain this information, we need to perform a decomposition on the Laplacian matrix to obtain eigenvectors and eigenvalues (the network spectrum). Sometimes, we normalize the Laplacian matrix first; in spectral clustering, this is less of a necessity than when computing specific graph theoretic values from the spectrum. Then, we find the Laplacian matrix's eigenvalues and eigenvectors. In practice, this is done through an algorithm rather than through linear algebra operations directly.

The second eigenvalue obtained from this process corresponds to the **Fiedler value** (also termed **algebraic connectivity**); this value calculates how many connected components exist in the network and relates to the robustness of the network. The corresponding eigenvector partitions the network into separate networks, thus providing a two-cluster solution to partitioning groups on the network. In spectral clustering, this process of partitioning a network into two separate networks can be repeated until a predefined stopping point is reached (such as a total number of clusters).

Spectral clustering on spatial networks constructed from ecological data allows researchers to partition out areas in which animals migrate, distinguish populations that might be isolated, and catalog differences in vegetation type or crop disease. Let's set up a simple example and explore different ways to compute spectral clustering.

Clustering ecological populations using spectral graph tools

Consider a researcher in Gabon hoping to understand the interconnectivity of Gaboon viper populations in a protected area of Loango National Park (*Figure 5.6*). Snakes are tagged in seven regions of the park (represented by a vertex), and their tags record their location for the next month. Snakes often will move between regions if their habitat overlaps with multiple regions. It's possible that multiple populations of snakes exist, giving different territories for different populations of snakes. If the two regions share a regional population, then the two vertices corresponding to these regions will have an edge. If no snakes migrate between regions, two vertices will not share an edge and likely represent different populations. Isolated populations tend to be more vulnerable to climate change, poaching, and habitat loss; understanding species and individual population ranges is critical to conservation efforts.

Figure 5.6 – An illustration of a Gaboon viper on the forest floor

Let's create some data on our snake population movements with `Script 5.1`:

```
#create snake area network adjacency matrix
import numpy as np
import networkx as nx
np.random.seed(0)

adj_mat = [[1,1,1,0,0,0,1],
           [1,1,1,0,0,0,0],
           [1,1,1,0,0,0,0],
           [0,0,0,1,1,1,0],
           [0,0,0,1,1,1,0],
           [0,0,0,1,1,1,1],
           [1,0,0,0,0,1,1]]

adj_mat = np.array(adj_mat)
```

In practice, geographic regions rarely share borders with all other regions in a dataset. **Spatial weighting** encodes **regional connectivity**. Geographically isolated areas are not likely to contain overlapping snake populations. However, it is possible that snakes will migrate across geographically isolated regions, and spatial weighting might not be appropriate in some animal population studies; however, Loango National Park does not include major barriers to travel like some mountainous or island geographies. Here, we'll add to `Script 5.1` spatial weighting component (weights total) to our adjacency matrix of snake overlap (`adj_mat`) to derive a weighted adjacency matrix (`adj_mat_w`):

```
#create spatial weight matrix
weights_total = [[1,1,1,0,0,0,1],
```

```
        [1,1,1,0,0,0,0],
        [1,1,1,0,0,0,0],
        [0,0,0,1,1,1,1],
        [0,0,0,1,1,1,1],
        [0,0,0,1,1,1,1],
        [1,0,0,1,1,1,1]]

weights_total = np.array(weights_total)
adj_mat_w=np.multiply(adj_mat,weights_total)
```

We can create and explore our degree matrix, as well as our Laplacian matrix (subtracting the adjacency matrix from the degree matrix). Let's add this information to Script 5.1:

```
#explore degree and Laplacian matrices
degree_matrix=np.diag(adj_mat_w.sum(axis=1))
laplacian_matrix=degree_matrix-adj_mat_w
print(degree_matrix)
print(laplacian_matrix)
```

The degree matrix encodes the number of vertices connected to a given vertex and serves as a basic measure of centrality. Hubs have high degree scores, while more isolated regions might have a degree of zero or one. Our network of snake population overlap has similar degree centrality measures across the entire network:

```
[[4 0 0 0 0 0 0]
 [0 3 0 0 0 0 0]
 [0 0 3 0 0 0 0]
 [0 0 0 3 0 0 0]
 [0 0 0 0 3 0 0]
 [0 0 0 0 0 4 0]
 [0 0 0 0 0 0 3]]
```

Let's now look at the Laplacian matrix, which encodes the connectivity information we'll use in our spectral decomposition and clustering:

```
[[ 3 -1 -1  0  0  0 -1]
 [-1  2 -1  0  0  0  0]
 [-1 -1  2  0  0  0  0]
 [ 0  0  0  2 -1 -1  0]
 [ 0  0  0 -1  2 -1  0]
 [ 0  0  0 -1 -1  3 -1]
 [-1  0  0  0  0 -1  2]]
```

There are multiple ways we can compute the Fielder vector to get our spectral clustering results. We'll go through two ways to compute this with Python. You may need to replace nx.from_numpy_matrix with nx.from_numpy_array, depending on your version of NetworkX. Depending on the size of the problem, it may be easier to use a linear algebra solver similar to the one we added to Script 5.1:

```
#define Fiedler vector and obtain clustering
G = nx.from_numpy_matrix(adj_mat_w)
ev = nx.linalg.algebraicconnectivity.fiedler_vector(G)
labels = [0 if v < 0 else 1 for v in ev]
labels

nx.draw(G,with_labels=True, node_color=labels)
```

This shows two distinct snake populations that converge to the sixth region of our Loango National Park sampling area (along with snakes staying in the exact same area, noted by loops), as shown in *Figure 5.7*:

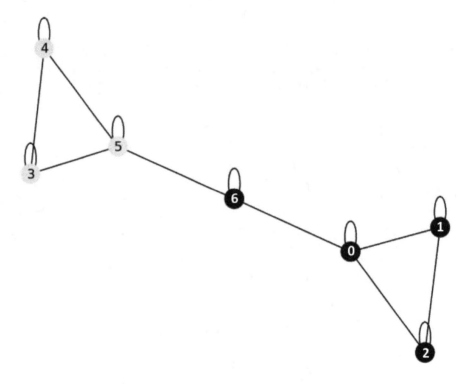

Figure 5.7 – Spectral clustering of snake populations

Scikit-learn provides a spectral clustering package, which we provide as an alternative calculation method by adding to Script 5.1. If you encounter a warning message on Windows, don't be alarmed; it is not an error. Let's add to Script 5.1 now:

```
#perform spectral clustering with sklearn
from sklearn.cluster import SpectralClustering
from sklearn import metrics
sc = SpectralClustering(2, affinity='precomputed', n_init=100)
sp_clust=sc.fit(adj_mat_w)
sc_labels=sp_clust.labels_
nx.draw(G,with_labels=True, node_color=sc_labels)
```

As expected, the scikit-learn package produces the same partitioning of the network (*Figure 5.8*) as our linear algebra approach:

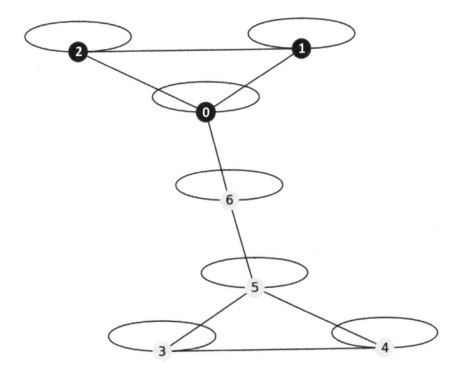

Figure 5.8 – Scikit-learn spectral clustering of snake populations

Spectral clustering is not limited to network data or ecological data. Any numeric dataset can be clustered using spectral clustering. In image and text analytics, we often don't have many instances of pre-labeled data, and manually annotating data is time consuming and can miss important classes in the dataset if they don't happen to be in a small sample for someone to score manually. Spectral clustering allows us to embed data through tools in these fields, create a network based on nearest neighbors, and apply spectral clustering to obtain labels for a dataset that can then be used to train a classification algorithm. This process is termed **semi-supervised learning**, and real-world data science in natural language processing and image analytics often relies on this methodology to build classification models from datasets. Let's see this in action by considering a text-based ecological dataset.

Spectral clustering on text notes

Let's consider a dataset of ecological notes, including five healthy and five unhealthy local parks. While these notes are not as extensive as a full ecological survey, they should demonstrate how to create a semi-supervised learning pipeline to classify ecosystem health. Here is one example of a healthy ecosystem in the notes:

```
"Mangrove ecosystem. Numerous butterflies, lizards, and insects.
Gumbolimbo trees. Soggy soil.', 'Tropical pines. Scat shows raccoon
and coyote populations here. Recent controlled burn."
```

Here is an example of an unhealthy ecosystem in the dataset:

```
"Small grass area. Some saplings. Gravel paths. Many cars parked in
non-parking zones. Lots of run-off."
```

While we could annotate these notes ourselves, we'll use spectral clustering to see how semi-supervised learning can be used to generate training data labels for much larger datasets where annotation might be unfeasible or unreliable. First, we'll import our data with `Script 5.2`:

```
#set up needed packages
import pandas as pd
import numpy as np
import os
#you may need to install sentence_transformers
#if the package is not already in your environment
#!pip install sentence_transformers
from sentence_transformers import SentenceTransformer


#import first network's data
File ="C:/users/njfar/OneDrive/Desktop/SC_Notes.csv"
pwd = os.getcwd()
os.chdir(os.path.dirname(File))
mydata = pd.read_csv(os.path.basename(File),encoding='latin1')
```

To embed our data, we'll use a **pretrained transformer model**—a type of deep learning algorithm trained on large sets of text data that is aware of language context and can map new text samples to the learned embedding space. While the particulars of this method are beyond the scope of this book, interested readers are encouraged to explore HuggingFace models and pretrained transformer models in the literature. Here, we use a pretrained HuggingFace model, a **Bidirectional Encoder Representations from Transformers (BERT)** model by first filling in missing data and then embedding it with the BERT model we've selected by adding to `Script 5.2`:

```
#prepare data
mydata['Notes']=mydata['Notes'].fillna(value=".")


#strip to text for input into BERT model
text_list=list(mydata.Notes)


#get BERT--768 vectors;
#note: you will need enough memory to load the transformer model
sbert_model1 = SentenceTransformer('all-mpnet-base-v2')
#encode data with BERT
encoded_text1=sbert_model1.encode(text_list)
```

Now that we have embedded our data, we can create our nearest neighbors graph. Since we have such a small sample size, we'll consider a point's three nearest neighbors by adding to `Script 5.2`:

```
#make nearest neighbors graph
from sklearn.neighbors import kneighbors_graph
n_adj = kneighbors_graph(encoded_text1, n_neighbors=3).toarray()
```

Now we're ready to run our spectral clustering and examine how our data clusters in this algorithm. Let's add to `Script 5.2` to see this in action:

```
#run spectral clustering
from sklearn.cluster import SpectralClustering
from sklearn import metrics
sc = SpectralClustering(2, affinity='precomputed', n_init=100)
sp_clust=sc.fit(n_adj)
sc_labels=sp_clust.labels_
print(sc_labels)
```

From our printed labels, we can see that ecosystems 3, 7, 8, and 10 form one cluster, while ecosystems 1, 2, 4, 5, 6, 7, and 9 form the other. Let's look at our sample of notes to see how well this method found differences in tone or theme by adding to `Script 5.2`:

```
#examine notes
print(text_list)
```

We can see that notes 3, 4, 7, 8, and 10 are unhealthy ecosystems compared to the rest of our sample. Spectral clustering doesn't seem to find all of these differences, but it's possible that another type of text embedding would distinguish tone better than the one we selected. It's also possible that fewer neighbors or a larger sample size would provide better results. However, we do find four of our five unhealthy ecosystems through our spectral clustering algorithm. A human reviewing results would not have to switch many labels (20%) to generate correct training labels for this data to be used to train a classification model. The second column of our dataset includes correct tone classification labels. Let's convert our BERT embeddings to a DataFrame by adding to `Script 5.2`:

```
#create training dataset for supervised learning
#turn BERT embedding into array
BERT_array1=np.array([x for x in encoded_text1])


#convert to dataframes
BERT_df1=pd.DataFrame(BERT_array1)
```

Now, let's split our data into training and test sets (80% and 20% splits, respectively), create a k-nearest neighbors classifier (which classifies points according to the labels of their two nearest neighbors with $k = 2$), and measure the test set error by adding to `Script 5.2`:

```
#create KNN classifier and test accuracy
from sklearn.neighbors import KNeighborsClassifier
from sklearn.model_selection import train_test_split

#get predictors and outcome
BERT_df1['Type']=mydata['Type']
df_train, df_test = train_test_split(BERT_df1,test_size=0.2, random_state=0)
X = df_train.iloc[:,0:767]
y = df_train.iloc[:,768]
X_test = df_test.iloc[:,0:767]
y_test = df_test.iloc[:,768]

#create KNN classifier and print accuracy
eu=KNeighborsClassifier(n_neighbors=2,metric='euclidean')
eu.fit(X,y)
print(eu.score(X_test,y_test))
```

Our split yields an accuracy of 100%, suggesting no misclassification of ecosystem types. Yours may vary given seeds and random number generators on your machine. However, this approach seems to work well for this dataset. Our pipeline of processing text data via embeddings, deriving a good start to label creation via spectral clustering, and then creating a classifier model yielded a high-accuracy classification system for our dataset.

Summary

In this chapter, we learned how to collect ecological data for a variety of data science problems. After a brief introduction to the theory of spectral clustering, we showed how spectral clustering could parse out different animal populations through our Gaboon viper distribution example. Finally, we explored spectral clustering of nearest neighbor networks that can be used in semi-supervised learning pipelines through an ecosystem note data example. In *Chapter 6*, we'll introduce centrality measurements and use them to find tipping points in stock pricing.

References

Angelici, F. M., Effah, C., Inyang, M. A., & Luiselli, L. (2000). *A preliminary radiotracking study of movements, activity patterns and habitat use of free-ranging Gaboon vipers, Bitis gabonica. Revue d'Ecologie, Terre et Vie, 55*(1), 45-55.

Corrales, D. C., Figueroa, A., Ledezma, A., & Corrales, J. C. (2015). *An empirical multi-classifier for coffee rust detection in Colombian crops. In Computational Science and Its Applications--ICCSA 2015: 15th International Conference, Banff, AB, Canada, June 22-25, 2015, Proceedings, Part I 15* (pp. 60-74). Springer International Publishing.

Froese, G. Z., Ebang Mbélé, A., Beirne, C., Atsame, L., Bayossa, C., Bazza, B., ... & Poulsen, J. R. (2022). *Coupling paraecology and hunter GPS self-follows to quantify village bushmeat hunting dynamics across the landscape scale. African Journal of Ecology, 60* (2), 229-249.

Mutombo, F. K. (2012). *Long-range interactions in complex networks.*

Ng, A., Jordan, M., & Weiss, Y. (2001). *On spectral clustering: Analysis and an algorithm. Advances in neural information processing systems, 14.*

Nunez-Mir, G. C., Iannone III, B. V., Pijanowski, B. C., Kong, N., & Fei, S. (2016). *Automated content analysis: addressing the big literature challenge in ecology and evolution. Methods in Ecology and Evolution, 7* (11), 1262-1272.

Qi, X., Fuller, E., Wu, Q., Wu, Y., & Zhang, C. Q. (2012). *Laplacian centrality: A new centrality measure for weighted networks. Information Sciences, 194,* 240-253.

Reimers, N., & Gurevych, I. (2019). *Sentence-bert: Sentence embeddings using siamese bert-networks. arXiv preprint arXiv:1908.10084.*

White, S., & Smyth, P. (2005, April). *A spectral clustering approach to finding communities in graphs.* In *Proceedings of the 2005 SIAM international conference on data mining* (pp. 274-285). Society for industrial and applied Mathematics.

Part 3:
Temporal Data Applications

Part 3 introduces temporal data applications, including problems related to stock market volatility prediction, change-point analysis of goods prices and sales volumes, and epidemic spread through dynamic animal interaction networks. In this part, we will get an overview of vertex- and edge-based centrality metrics, extensions of network data structures to simplicial complexes, triadic closure, and dynamic epidemic models.

Part 3 has the following chapters:

- *Chapter 6, Stock Market Data*
- *Chapter 7, Goods Prices/Sales Data*
- *Chapter 8, Dynamic Social Networks*

6

Stock Market Data

In this chapter, we'll introduce **temporal data** and dive into stock market trend analysis. To understand trends over time, we'll return to **centrality measurements** on networks and introduce some more advanced algorithms. Finally, we'll analyze stock pricing data over time using our centrality measurements and pinpoint changes in behavior over time within and across different stocks to predict spikes and crashes in price.

By the end of this chapter, you'll be able to wrangle datasets with time components into a series of networks and analyze structural changes over time with centrality metrics. Many of the centrality metrics scale well to large networks, particularly when they are run in parallel.

Specifically, we will cover the following topics:

- Introduction to temporal data

- Introduction to centrality metrics

- Application of centrality metrics across time slices

- Extending network metrics for time series analytics

Let's get started by returning to temporal datasets and the limitations of non-network-based models to analyze them.

Technical requirements

You will require Jupyter Notebook to run the practical examples in this chapter.

The code for this chapter is available here: `https://github.com/PacktPublishing/Modern-Graph-Theory-Algorithms-with-Python`

Introduction to temporal data

In *Chapter 2*, we briefly introduced temporal data or data in the form of a time series or a group of time series. **Time series data** tracks important metrics in many different industries: daily store sales volumes, weekly software product marketing lead volumes, daily incidence of an emerging disease, yearly behavior rates (such as smoking or vegetable consumption) in a population, or hourly stock prices tracking market trends. Many related factors can influence trends over time, and some models consider these factors directly if they are known in advance.

However, consider the case of sales trends for a new gem store in a city where gem stores are a new phenomenon, perhaps somewhere rural between Haifa and Tel Aviv (*Figure 6.1*). Thus, there is very little known about what might influence sales. Understanding what trends exist in the time series data is critical when mining for factors that might influence sales. However, time series datasets pose significant challenges to many supervised learning methods, such as **random forest models** or **linear regression**. At one point in time, sales are not independent; they rely on factors that influence sales in the previous days, limiting the use of supervised learning and many types of **unsupervised learning**. The lack of predictors also poses a challenge.

Figure 6.1 – An illustration of a gem store located partway between Tel Aviv and Haifa, Israel

Fortunately, for our gem store, many algorithms are designed to handle time series data, such as **autoregressive integrated moving average models (ARIMA models)**, **singular spectrum analysis (SSA)**, and **Holt–Winters models**. However, changes in time series behavior (spikes, crashes, and changes in variance) pose a challenge to these models. Capturing and predicting these changes is critical if you want to create a predictive model or mine for factors influencing the time series values. In our gem store example, seasonality in tourism, conflicts in the region, and holiday travel promotions may influence traffic along the route in which our store is positioned.

One industry with abundant and very complex time series data is finance. Many social and economic factors influence stock prices, and untangling the relationships and randomness in stock price fluctuations underlies much of the finance industry. Let's have a look at the stock market pricing data and common tasks in analyzing stock market pricing data.

Stock market applications

In recent years, the financial sector has shifted from an expert-driven model of stock market insight to a more machine learning-based approach. Machine learning models sift out emerging trends and catch subtle trends that may escape a human pouring over the data. This tactic also allows analysts to process a much larger data collection than a human could process, including many different sectors, international stock exchanges, and even individual frontier markets. By collecting more insight, it is possible for investors and investment management firms to invest in a wider variety of markets without as much expertise in those markets.

Stock market analytics covers a vast field of applications. Investments can be made within certain market sectors (such as technology or agriculture) or across markets. They can focus on short-term gains (including those made in minutes or hours) and long-term gains (which may span decades). They can also focus on foreign markets, where stock prices may be influenced by factors very different from those influencing a local market's prices. All these scenarios guide analytics efforts and the time scale of data collected for analysis.

In general, analyzing stock market data involves assessing many types of trends over time. Stocks can have constant prices over time, experience gradual price growth and reduction, crash suddenly, or grow exponentially. Each of these suggests a different purchase/sale strategy for investors in the short term and the long term. *Figure 6.2* shows a hypothetical stock that exhibits many of these patterns:

Figure 6.2 – An example of stock data trends, including stagnant periods, growth, shrinkage, and a crash

We can see in *Figure 6.2* that Stock A begins 2022 with a consistent price before entering a growth phase around July 2022. This growth phase lasts until early 2023, at which point it enters a constant pricing phase again. As an event happens in March of 2024, the price of Stock A crashes and then enters a period of price decline until the end of our tracked time period.

Often, these trends do not occur in isolation. The stocks of companies that share supply chains may exhibit similar trends. Stocks in the same industry may exhibit similar or opposite trends, depending on how companies relate to each other or news in the industry. Stocks in shared trade or defense regions may exhibit similar trends, as well, given the sociopolitical ties across countries and their markets (such as the COVID crash across most economies).

Tipping points, where trends change dramatically, attract a lot of attention in financial analytics. These represent opportunities to invest before a period of accelerating growth or warnings to pull out of a market or particular investment before a crash. However, detecting these trends challenges many of the commonly used tools in market analytics.

Thankfully, newer tools, including a few rooted in network science, identify tipping points more readily than traditional methods. Many tools hinge on large-scale coupling across markets, sectors, or stocks. As more and more stocks (or markets) exhibit similar behavior, the system becomes vulnerable to outside influences that can tip it into a crash (such as supply chain issues, new legislation, or a pandemic). Simply calculating the correlations among individual stocks or sectors of a market can provide some insight, but transforming correlations (within slices of time) to networks allows us to leverage many network science tools that dive deeper into the nature of their correlations and their changes over time. Specifically, centrality metrics allow us to quantify and classify relationships that exist within a network. Let's explore a few of these centrality metrics.

Introduction to centrality metrics

We've encountered some centrality metrics in *Chapter 3*, where we learned about bridges and hubs. Many vertex-based centrality metrics calculate properties related to hubs—the connection of a vertex to its nearest neighbors and their nearest neighbors. Many edge-based centrality metrics calculate bridging properties, where the edges near a vertex act as connectors between different hubs.

Degree is the simplest **vertex-based centrality metric**, which we encountered in *Chapter 5*. **Degree centrality** is simply the number of vertices directly connected to the vertex of interest. Many Laplacian-based metrics or algorithms depend on the degree matrix within algorithm calculations. On the surface, this metric seems to capture important hub properties; a vertex with a high degree centrality will carry a lot of influence within the network (and, thus, might make a good intervention target). It also scales well to very large networks. However, one limitation of degree centrality is its lack of awareness of a vertex's position beyond any immediate connections to neighbors; a vertex with a low degree centrality may be connected to many vertices with a high degree centrality, giving it more influence over network behavior and structure than its degree centrality suggests.

Eigenvector centrality and its variants (including **PageRank** and **Katz centrality**) incorporate awareness about connectivity beyond immediate neighbors to give a more comprehensive vertex-based centrality metric. Thus, **eigenvector centrality** would score our hypothetical low-degree centrality vertex connected to high-degree centrality vertices highly, as that vertex is near very connected vertices. Technically, to find the eigenvector centrality of each vertex in a network, we can perform an eigen decomposition on the adjacency matrix. Because the adjacency matrix does not include negative values, we can assume that the first eigenvalue is the largest, and its eigenvector yields the eigenvector centrality scores for our vertices.

PageRank centrality extends eigenvector centrality and increases its flexibility by replacing the adjacency matrix with an adjusted adjacency matrix that is constructed by performing a random walk across vertices to determine the adjacency properties. In addition, random surfer properties, where a random walk can cross the unconnected areas of a graph with a low probability, create an adjusted adjacency matrix that is connected. This adjusted adjacency matrix is then scaled before performing the eigen decomposition on the adjusted adjacency matrix, which is carried out to compute eigenvector centrality scores. PageRank centrality scores, thus, provide flexibility. In addition, this computation is typically easier and faster with the adjusted adjacency matrix, allowing the algorithm to scale well to networks of even hundreds of millions of vertices.

Edge-based centrality measures capture network infrastructure that is important for spreading processes and connectivity across different hubs. **Betweenness centrality** is one of the most common edge-based centrality metrics, capturing the relative number of shortest paths that include a given vertex among all shortest paths that exist in the network. Consider a network with 10 shortest paths, 8 of which include a particular vertex. Without this vertex, many of the shortest paths would not exist, inconveniencing the network greatly in terms of spreading processes on a social network or route efficiency on a transportation network. However, betweenness centrality does not scale well, and it should not be used on large networks without some sort of parallelization of the operation.

One recent edge-based centrality network has proven to be an efficient tool for finding stock market tipping points. **Forman–Ricci curvature** is a geometry-based tool that considers adjacent edges in relation to an edge of interest. On an unweighted network, Forman–Ricci curvature is calculated by subtracting the degree centrality of the two vertices attached to an edge of interest from 2. The constant, 2, represents the connection between the two vertices connected by the edge of interest. The degree centrality of both vertices connected by the edge counts the number of adjacent edges to our edge of interest (minus the vertices connected by that edge, whereby we obtain our constant of 2). In *Figures 6–3*, we see three vertices: two vertices with a single edge in relation to the middle vertex (both with a degree centrality of 1) and a middle vertex connecting to both of the outer vertices (with a degree centrality of 2). Both edges, thus, have a Forman–Ricci curvature of -1, as the sum of the vertex degree centralities for both is 3 (2 - 3 = -1).

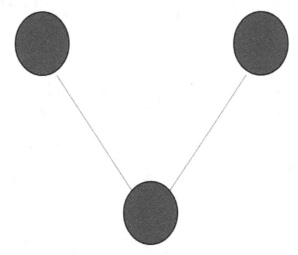

Figure 6.3 – An example network demonstrating Forman–Ricci curvature, where the middle vertex is pulled from both the first and third outer vertices

To obtain vertex centrality metrics from this edge metric, we can sum the edge metrics for each vertex to score the vertices. In our example in *Figure 6.3*, we have outer vertices with a Forman–Ricci vertex centrality of -1, as both only connect to a single edge. However, the middle vertex connects to two edges, with a Forman–Ricci curvature of -1, giving it a Forman–Ricci vertex centrality of -2. Because this centrality metric relies on low-cost computations, it scales well to large networks and can be used as an alternative edge-based centrality score when betweenness centrality is not feasible.

Now, let's explore some stock market data and see how network science can help us identify key trends over time.

Application of centrality metrics across time slices

The NASDAQ stock market is an American stock exchange in New York City that includes publicly traded companies such as Apple, Alphabet, Nvidia, and Microsoft. Kaggle provides a full history of NASDAQ stock prices under a public license (`https://www.kaggle.com/datasets/jacksoncrow/stock-market-dataset?resource=download`) that can give us stock data for these four tech companies during the period in which they were all publicly traded up to April 1, 2020. We've munged the data for you to include only these four stocks in the period from August 19, 2004, to April 1, 2020. Let's take a peek at these data to see what trends might exist (*Figure 6.4*):

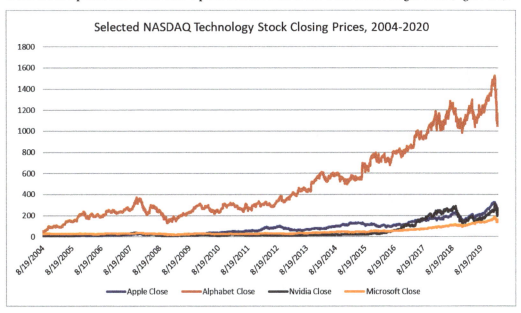

Figure 6.4 – NASDAQ selected stock closing values from August 19, 2004, to April 1, 2020

In *Figure 6.4*, we can see that Alphabet has a consistently higher price, but all stocks exhibit the typical trends of constant pricing, dips, spikes, and upward or downward trends over this long period of trading. Of note is the 2020 trend, where stocks experienced the COVID-19 crash. Periods of volatility include 2008–2009 and 2016–2020.

One of the important aspects of time series analysis is **windowing** the time series data into overlapping pieces. While windowing can be optimized by using a grid search, windowing tends to be more of an art than a science. Choosing a window impacts the length of time in which trends can be captured. In our stock market data, window length limits the period in which we can search for trends and, thus, limits the time period after the last window in which we can forecast market behavior. A window that is too large can miss important trends that impact short-term market behavior. A window that is too small can limit forecasting to the immediate future.

We'll choose a window of 5 trading days or roughly a week's worth of stock data. This allows us to capture trends relevant to day trading, where stocks are traded frequently based on volatility. Our network metrics should work well for volatility-based trading for quick gains, as they capture increasing correlations across the two stocks.

Another aspect that is important to windowing time series data is the choice of overlap. For our example, we'll choose maximal overlap (4 days' overlap) to maximize our sensitivity to day-to-day trends. In other applications, less overlap may be desirable to investigate longer time trends. Note that our example uses a path on one of our machines. Your file path will be different from ours.

Let's import our packages and load our data into Python with `Script 6.1` to get started:

```
#import packages
import igraph as ig
from igraph import Graph
import numpy as np
import pandas as pd
import os
import matplotlib as plt

#import stock data
File="C:/users/njfar/OneDrive/Desktop/AAPL_GOOGL_Stock_2004_2020.csv"
pwd=os.getcwd()
os.chdir(os.path.dirname(File))
mydata=pd.read_csv(os.path.basename(File),encoding='latin1')
```

Now that we have our data imported, we can add a loop that windows our time series to 5-day periods that overlap by 4 days across slices. This window strategy yields the best chance to find daily changes in trends. We'll then create correlations among the four stocks within that time slice, threshold those correlations to limit our analysis to high correlations, create an unweighted network from those thresholds, and analyze Pagerank centrality, degree centrality, betweenness centrality, and Forman–Ricci curvature centrality across the time slices. We'll also save each unweighted network for future retrieval, as well as the network metrics and their averages over each time slice. Let's add these pieces to `Script 6.1`:

```
#script to create time slices, derive networks,
#and compute centrality metrics
stock_networks=[]
bet_t=[]
deg_t=[]
eig_t=[]
vcurv_t=[]
bet_ave=[]
deg_ave=[]
```

```
eig_ave=[]
vcurv_ave=[]

for Date in range(5,3932):
    #wrangle data into graph
    data=mydata.iloc[(Date-5):(Date),1:5]
    cor=np.corrcoef(data.transpose())
    cor[cor>=0.5]=1
    cor[cor<0.5]=0
    stock_data=Graph.Adjacency(cor)
    stock_networks.append(stock_data)
    #derive some centrality metrics
    d=Graph.degree(stock_data)
    deg_t.append(d)
    deg_ave.append(np.mean(d))
    b=Graph.betweenness(stock_data)
    bet_t.append(b)
    bet_ave.append(np.mean(b))
    e=Graph.pagerank(stock_data)
    eig_t.append(e)
    eig_ave.append(np.mean(e))
    #create Forman-Ricci curvature calculations
    ecurvw=[]

    for edge in stock_data.es:
        s=edge.source
        t=edge.target
        ecurvw.append(2-d[s]-d[t])
    vcurvw=[]

    for vertex in stock_data.vs:
        inc=Graph.incident(stock_data,vertex)
        inc_curv=[]
        for i in inc:
            inc_curv.append(ecurvw[i])
        vcurvw.append(sum(inc_curv))
    vcurv_t.append(vcurvw)
    vcurv_ave.append(np.mean(vcurvw))
```

This script should run quickly on your machine. If you include a very large number of stocks in your analyses, you may wish to run the script in parallel to save time or exclude betweenness centrality, as betweenness centrality does not scale well as the number of vertices increases.

Now that we have computed our metrics, let's examine the correlations between the metrics across our set of time series windows to see how the different metrics relate to each other. Given that Forman–Ricci curvature depends on degree centrality metrics, we'd expect to see a strong correlation. We can add to Script 6.1 to obtain these correlations:

```
#examine correlations among metrics across the time series
print(np.corrcoef(deg_ave,eig_ave))
print(np.corrcoef(deg_ave,bet_ave))
print(np.corrcoef(deg_ave,vcurv_ave))
print(np.corrcoef(eig_ave,bet_ave))
print(np.corrcoef(eig_ave,vcurv_ave))
print(np.corrcoef(bet_ave,vcurv_ave))
```

You should see a strong correlation between degree centrality and Forman–Ricci curvature centrality (-0.99 in our analysis) but fairly weak correlations among the other centrality metrics (-0.05 for degree and Pagerank centralities, 0.01 for degree and betweenness centralities, 0.04 for Pagerank and betweenness centralities, 0.04 for Pagerank and Forman–Ricci curvature centrality, and 0.05 for betweenness and Forman–Ricci curvature centrality). This suggests that degree centrality and Forman–Ricci curvature may be interchangeable in these analyses, though a more complex Forman–Ricci curvature metric may capture a bit more information that may be relevant to certain trends. It's unclear if these correlations would hold for other datasets, though degree centrality and Forman–Ricci curvature on unweighted networks will correlate to some extent, given that the Forman–Ricci curvature formula depends on degree centrality.

Let's visualize the network metric trends over time to see if any patterns emerge or extreme values stand out that might indicate behavior changes in our stock pricing. We'll add visualization code to Script 6.1 to do this:

```
#plot metric averages across time slices
time=range(0,3927)
plt.plot(time, deg_ave, label = "Degree Average")
plt.plot(time, eig_ave, label = "Pagerank Average")
plt.plot(time, bet_ave, label = "Betweenness Average")
plt.plot(time, vcurv_ave, label = "Forman-Ricci Curvature Average")
plt.legend()
plt.show()
```

This should give you a plot similar to *Figure 6.5*:

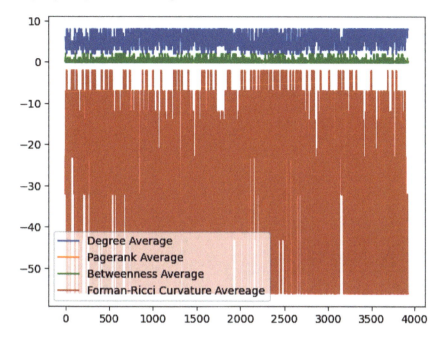

Figure 6.5 – A plot of centrality averages across time slices for our stock market data

Note that Pagerank centrality does not show up in our plot. Pagerank and betweenness centrality fill a similar range of values, masking Pagerank centrality in our plot. However, we do see that differences in the average centrality values emerge regularly in our plot, suggesting that our centrality values may be capturing different states of the market over time.

Our Forman–Ricci curvature centrality averages suggest periods of relative stability in the market, where the values are near zero. Two prominent periods of relative stability occur at the start of our time series (roughly 2004–2006) and again in the mid-to-late 2010s. However, as we approach 2008 and 2020, the correlations among our stocks increase considerably before two major market crashes.

It's probable that our choice of threshold value influences our results. To hone in on periods of tight coupling in terms of stock behavior, let's raise our threshold value to `0.9` and rerun `Script 6.1`:

```
#script to create time slices, derive networks,
#and compute centrality metrics
stock_networks=[]
bet_t=[]
deg_t=[]
eig_t=[]
vcurv_t=[]
```

```
bet_ave=[]
deg_ave=[]
eig_ave=[]
vcurv_ave=[]

for Date in range(5,3932):
    #wrangle data into graph
    data=mydata.iloc[(Date-5):(Date),1:5]
    cor=np.corrcoef(data.transpose())
    cor[cor>=0.9]=1
    cor[cor<0.9]=0
    stock_data=Graph.Adjacency(cor)
    stock_networks.append(stock_data)
    #derive some centrality metrics
    d=Graph.degree(stock_data)
    deg_t.append(d)
    deg_ave.append(np.mean(d))
    b=Graph.betweenness(stock_data)
    bet_t.append(b)
    bet_ave.append(np.mean(b))
    e=Graph.pagerank(stock_data)
    eig_t.append(e)
    eig_ave.append(np.mean(e))

    #create Forman-Ricci curvature calculations
    ecurvw=[]
    for edge in stock_data.es:
        s=edge.source
        t=edge.target
        ecurvw.append(2-d[s]-d[t])
    vcurvw=[]

    for vertex in stock_data.vs:
        inc=Graph.incident(stock_data,vertex)
        inc_curv=[]
        for i in inc:
            inc_curv.append(ecurvw[i])
        vcurvw.append(sum(inc_curv))
    vcurv_t.append(vcurvw)
    vcurv_ave.append(np.mean(vcurvw))
```

Now, we can rerun our correlation analysis to understand how threshold value might influence any correlations among the metrics. Let's rerun our correlation analysis in Script 6.1:

```
#examine correlations among metrics across the time series
print(np.corrcoef(deg_ave,eig_ave))
print(np.corrcoef(deg_ave,bet_ave))
print(np.corrcoef(deg_ave,vcurv_ave))
print(np.corrcoef(eig_ave,bet_ave))
print(np.corrcoef(eig_ave,vcurv_ave))
print(np.corrcoef(bet_ave,vcurv_ave))
```

You should see some notable differences compared to our prior results with this new threshold. The degree and Pagerank centralities are still not correlated very much (-0.04), which is mirrored by the correlations between Pagerank centrality and betweenness centrality (0.003) and Pagerank and Forman–Ricci curvature centrality (0.04). However, degree and betweenness centrality are moderately correlated now (0.44), as are betweenness and Forman–Ricci curvature centrality (-0.39). The degree and Forman–Ricci curvature centralities are still highly correlated, though slightly less than in our prior threshold (-0.98).

Let's plot our new results to investigate any trends that may have been masked with a low threshold value in our initial analysis. We can replot this using Script 6.1:

```
#plot metric averages across time slices
time=range(0,3927)
plt.plot(time, deg_ave, label = "Degree Average")
plt.plot(time, eig_ave, label = "Pagerank Average")
plt.plot(time, bet_ave, label = "Betweenness Average")
plt.plot(time, vcurv_ave, label = "Forman-Ricci Curvature Average")
plt.legend()
plt.show()
```

This should yield a plot that looks very different from *Figure 6.5*. In *Figure 6.6*, we can see the periods of volatility much more clearly than we could in *Figure 6.5*:

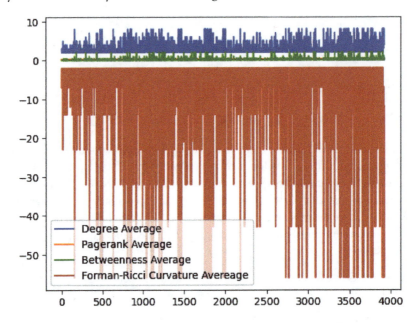

Figure 6.6 – A plot of network centrality metrics across time slices with a 0.9 threshold

Figure 6.6 shows many more Forman–Ricci curvature values that are near 0, suggesting less volatility in the market during those time periods. We see a few periods of increasing volatility (more extreme Forman–Ricci curvature values) in the form of dips in our plot. Those periods of intense volatility and coupling precede the crashes in 2008 and 2020, as well as periods of quick rebuilding after crashes. These are periods of interest for investors, as large sums of money are either lost or gained during those periods.

In general, network metrics seem to pick up on market volatility very well, particularly when high thresholds are applied to the data. This suggests what has been stated in the recent literature: network metrics are useful tools to identify market volatility before crashes and exponential growth. In fact, these tools picked up growing market volatility long before the COVID crash of 2020, which could have saved gains in the months leading up to 2020 had investors heeded the volatility warnings and pulled out before trouble hit the market. Given the volatility and long period of steep growth, it's likely that any number of factors would have caused a major crash. A large-scale or badly placed regional conflict, a breakdown in the supply chain, or a change in policy within the technology sector probably would have produced a major crash.

Returning to *Figure 6.7*, we observe very different trends before the 2008 crash than the 2020 crash. While 2020 was preceded by a long period of growth with recent small crashes, 2008 was preceded by relative stability and slow growth. Given the increasing volatility and accelerated growth of stock prices for these four NASDAQ stocks, these trends make the market more vulnerable to large crashes in the future, and should the trends hold across other NASDAQ sectors, then we'd expect less certainty and more opportunities for large gains and losses in the near future.

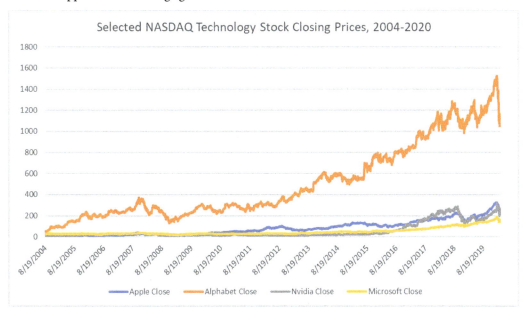

Figure 6.7 – Returning to the plot of the stock data

Given these trends in the market, network science tools are poised to play a critical role in stock market analytics. Relatively little work exists in terms of the application of these tools to stock market data or time series data more generally, and few centrality metrics have been studied systematically. Much of the existing research on the application of centrality metrics to understand stock market trends over time involves extensions of networks to include multi-way relationships. Let's now turn our attention to an extension of applying networks to the relationships that exist among more than two entities.

Extending network metrics for time series analytics

Because networks are topological objects and because our correlation matrix can use a threshold of any value, it is possible to extend network metrics to the realm of topological data analysis. Networks capture the two-way relationships between entities. However, three-way, four-way, and even larger-way interactions can exist, as well. **Simplicial complexes** extend the idea of networks to capture these higher-numbered interactions. Three-way interactions are represented as faces (or triangles) outlined by two-way lines. Four-way interactions are represented as tetrahedra, comprised of three-way faces that have four-way interactions. This process can continue to any value of mutual interactions, where the lower-dimensional interactions form the boundaries of the higher-dimensional interactions. A simplicial complex collects the highest-level interactions that exist among vertices into a single object. *Figure 6.8* shows an example of a three-way interaction bounded by mutual, two-way interactions:

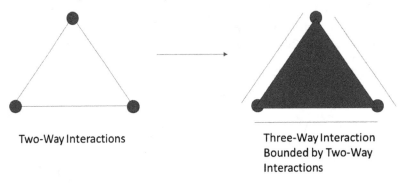

Two-Way Interactions

Three-Way Interaction
Bounded by Two-Way
Interactions

Figure 6.8 – A diagram showing how a three-way interaction is defined by mutual, two-way interactions

Just as networks can be weighted or unweighted, simplicial complexes can be weighted or unweighted across n-way interactions within the simplicial complex. In addition, just as networks are constructed from an adjacency matrix, so simplicial complexes are created from adjacency matrices at each level of n-way interactions (technically the boundary matrices). Just as we typically need to construct an adjacency matrix to create a network from raw data, we can use raw data to define the n-way interactions existing in that data.

First, a relationship metric must be defined for the raw data. In our stock dataset, we chose correlation metrics. Distance metrics are also commonly used when constructing both networks and simplicial complexes, and many, many distance metrics can be defined for continuous or discrete measurements on a dataset. Once a metric is defined, a threshold or series of thresholds are applied to the metric matrix to define the relationships within a given radius of each other (defined by the threshold). Two main options exist for constructing the simplicial complex and its relationships:

- Defining the n-way relationships through the union of points, the radii of which touch (called a **Vietoris–Rips complex**)

- Counting the connection points, which involves the mutual overlap of points within a given radius, where all connected points must lie within each other's radius (called a **Čech complex**)

The **filtration** of simplicial complexes involves varying the radius by different metric thresholds. Remember how applying different correlation thresholds produced different results and insights for our stock market dataset? Different filtration levels of simplicial complexes can produce different simplicial complexes at each filtration level with different properties that may contain important information for an analyst.

Let's create a function that defines the Vietoris–Rips simplicial complex for two-way interactions (corresponding to a network) using `Script 6.2`:

```
#define Vietoris-Rips complex
from itertools import combinations
from numpy import linalg as LA
def graph_VR(points, eps):
    points=[np.array(x) for x in points]
    vr=[(x,y) for (x,y) in combinations(points, 2)
    if LA.norm(x - y) <= 2*eps]
    return np.array(vr)
```

Now, we can apply this to the first slice of our stock market time series data. We'll choose thresholds of 1 and 10 as a starting point to understand which vertex pairs will be included in our simplicial complex (here, only at the two-way interaction level). Let's add this to `Script 6.2` to calculate the Vietoris–Rips simplicial complex for the first slice of our stock market time series data:

```
#apply Vietoris-Rips with multiple thresholds to a slice of our stock
#dataset
data=mydata.iloc[0:5,1:5]
vr1=graph_VR(data.transpose(),1)
vr2=graph_VR(data.transpose(),10)
```

We can examine the vertex pairs included in each filtration by adding the following to `Script 6.2`:

```
#print the results
print("Vietoris-Rips Complex, Threshold=1:")
print(vr1)
print("Vietoris-Rips Complex, Threshold=10:")
print(vr2)
```

This gives us pairs of vertices for each filtration, which should show the following:

```
Vietoris-Rips Complex, Threshold=1:
[[0 1]
 [0 2]
 [1 2]
 [1 3]
 [2 3]
 [2 4]
 [3 4]]
Vietoris-Rips Complex, Threshold=10:
[[0 1]
 [0 2]
 [0 3]
```

```
     [0 4]
     [1 2]
     [1 3]
     [1 4]
     [2 3]
     [2 4]
     [3 4]]
```

We can visualize the threshold = 1 complex with two-way interactions by adding the following to Script 6.2:

```
edges1 = [(0,1),(0,2),(1,2),(1,3),(2,3),(2,4), (3,4)]
import networkx as nx
G1 = nx.Graph()
G1.add_edges_from(edges1)
nx.draw(G1,with_labels=True)
```

This gives a figure similar to the one in *Figure 6.9*, showing a set of three triangles connected by two-way interactions:

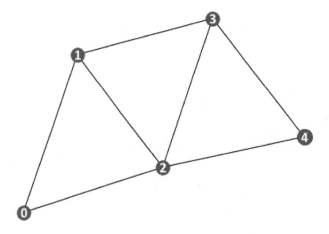

Figure 6.9 – A visualization of the threshold = 1 results for two-way interactions

Similarly, we can visualize our threshold = 10 results by adding the following to Script 6.2:

```
edges10 = [(0,1),(0,2),(0,3),(0,4),(1,2),(1,3),(1,4),(2,3),(2,4),(3,4)]
import networkx as nx
G10 = nx.Graph()
G10.add_edges_from(edges10)
nx.draw(G10,with_labels=True)
```

This should give a plot similar to *Figure 6.10*, which shows a more complex connectivity than the threshold = 1 results:

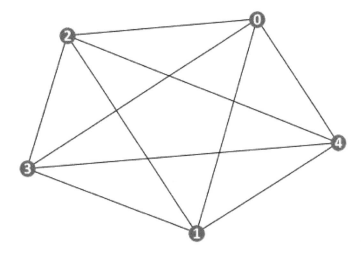

Figure 6.10 – A visualization of the threshold = 10 results for two-way interactions

These results show that different radius thresholds yield different networks, just as our correlation thresholds produced different networks. We could build a full filtration from the first appearance of an edge until all possible edges exist in our network to track changes in network structure based on distances between points. We could also include three-way and four-way interactions in our construction of the Vietoris–Rips complex to extend our analysis to the fully simplicial complexes that exist in each slice of our dataset.

While simplicial complex metrics are beyond the scope of this book, many extensions of network metrics to simplicial complexes exist (such as Forman–Ricci curvature centrality) and may merit investigation in the analysis of time series data. Currently, very little work has elucidated the use of simplicial complex metrics or methods on stock market change point detection.

If you are interested, we encourage you to extend `Scripts 6.1` and `6.2` to include an analysis of the full time series through the lens of simplicial complexes and calculate the extensions of the network metrics in terms of simplicial complexes. Our chapter reference section includes papers that experiment with these extensions.

Summary

In this chapter, we've introduced some common uses of time series data and time series analytics, including stock market data. We explored several vertex- and edge-based centrality metrics that are common in network analytics. Then, we applied network metrics to a time series problem on NASDAQ stock data from 2004 to 2020 to investigate how network metrics and time series thresholding impact the ability to extract useful information from time series data, such as our stock data. Finally, we investigated extending networks to simplicial complexes and constructed a network by building two simplicial complexes using the Vietoris–Rips method and various threshold values. In *Chapter 7*, we'll look at sales and goods pricing across both time and geography to see how network science can solve problems in spatiotemporal data.

References

De Floriani, L., & Hui, A. (2005, July). *Data Structures for Simplicial Complexes: An Analysis And A Comparison*. In Symposium on Geometry Processing (pp. 119-128).

Durbach, I., Katshunga, D., & Parker, H. (2013). *Community structure and centrality effects in the South African company network. South African Journal of Business Management*, 44(2), 35-43.

Estrada, E., & Ross, G. J. (2018). *Centralities in simplicial complexes. Applications to protein interaction networks. Journal of theoretical biology*, 438, 46-60.

Johansen, A., & Sornette, D. (1998). *Stock market crashes are outliers. The European Physical Journal B-Condensed Matter and Complex Systems*, 1, 141-143.

Rodrigues, F. A. (2019). *Network centrality: an introduction. A mathematical modeling approach from nonlinear dynamics to complex systems*, 177-196.

Salnikov, V., Cassese, D., & Lambiotte, R. (2018). *Simplicial complexes and complex systems. European Journal of Physics*, 40(1), 014001.

Samal, A., Pharasi, H. K., Ramaia, S. J., Kannan, H., Saucan, E., Jost, J., & Chakraborti, A. (2021). *Network geometry and market instability. Royal Society open science*, 8(2), 201734.

Valente, T. W., Coronges, K., Lakon, C., & Costenbader, E. (2008). *How correlated are network centrality measures? Connections* (Toronto, Ont.), 28(1), 16.

Xiong, J., & Xiao, W. (2021). *Identification of key nodes in abnormal fund trading network based on improved pagerank algorithm*. In *Journal of Physics*: Conference Series (Vol. 1774, No. 1, p. 012001). IOP Publishing.

Goods Prices/Sales Data

In this chapter, we'll combine spatial and temporal approaches to **spatiotemporal data**—data that includes both spatial components and time series components. To handle the time components, we'll slice our datasets into overlapping time windows as we did with our stock data in *Chapter 6*. To handle the spatial components, we'll calculate local Moran statistics based on correlations for each time slice and threshold the value to create a network for that time slice. Then, we'll look at changes in **Forman-Ricci curvature centrality** and **PageRank centrality** across slices of time and space. Examples include the Burkina Faso millet dataset first seen in *Chapter 2* and a new store sales dataset.

By the end of this chapter, you'll understand how to set up spatiotemporal data analytics with networks to capture changes over time within the spatial and temporal relationships of the datasets using **igraph** and different types of plots to look at spatial relationships within slices of time.

In particular, we will cover the following topics:

- An introduction to spatiotemporal data
- Analyzing our spatiotemporal data

Technical requirements

You will require Jupyter Notebook to run the practical examples in this chapter.

The code for this chapter is available here: `https://github.com/PacktPublishing/Modern-Graph-Theory-Algorithms-with-Python`

An introduction to spatiotemporal data

In *Chapter 2*, we discussed some aspects of spatial and temporal data that make such data difficult to wrangle, analyze, and scale with statistical and machine learning methods. Recall that both types of data are not independent, with dependencies crossing either space or time in the dataset. Alternative approaches to prediction are needed, and representations of spatial and temporal data as a network allow analysts to leverage the tools of network science to find change points, mine for patterns, and even predict future system behavior in the case of temporal data. In the previous chapters, we've learned how to analyze both spatial and temporal data with network science tools such as centrality to understand important trends.

Many analytic tasks in the real world involve spatiotemporal data—including the analysis of world stock markets influenced by local policies, data mining of store data across locations for trends in customer behavior and revenue, discerning disease sources and population vulnerabilities during an epidemic, tracking human behavior (such as crime, eating habits, substance use, birth rates, or terrorism) across countries or regions of a specific country, and demonstrating the effects of climate change across vulnerable countries.

Let's get acquainted with our two datasets and problems relevant to many applications of spatiotemporal data.

The Burkina Faso market dataset

In *Chapter 2*, we introduced the Burkina Faso market dataset, which recorded market millet prices across the country from Quarter 2 of 2015 to Quarter 2 of 2022. During this period, world grain prices fluctuated due to COVID-19 supply chain issues and the Ukraine conflict limiting food distribution and aid. Food security is a major issue in many regions of the world, and understanding what might cause prices to soar out of reach for the average person or family is critical in meeting the world's food needs and avoiding famine.

Some regions of a country may be more vulnerable to food shortages or soaring costs of goods than others, and the spatial features in our dataset capture these for the Burkina Faso markets. We'd imagine a rural market (such as the one shown in *Figure 7.1*) would face greater pricing and availability instability than a market in the heart of a city such as Ouagadougou. In addition, some events within a period may cause more pricing volatility than other events in that same period. Capturing the time series trends is also important to understand the causes and potential effects of events on grain prices in the region. Thus, capturing the data as spatiotemporal data allows us to analyze both trends across locations and across time.

Figure 7.1 – An illustration of a rural market in Burkina Faso

To capture spatial trends, we'll return to the weight matrix introduced in *Chapter 2*, which connected markets located in provinces that share a border. This roughly captures the geography of Burkina Faso, allowing regions that share many borders the potential to influence each other's pricing and demonstrating the potential supply chain issues related to remote regions of the country.

Just as pricing can change over time and geography, so can sales volumes. Let's turn to our next example and explore how sales can change across stores due to COVID-19 changes in work policies and the movements of populations across a city.

Store sales data

Let's consider a local chain of printer supply stores in a medium-sized city. As populations change near store locations, it may be necessary to pivot business strategy: closing stores that are struggling to minimize real estate costs, preferentially stocking certain stores that are doing well, and opening new locations in high-volume sales areas. Perhaps work-from-home initiatives during COVID-19 have resulted in more people working from home in certain areas of the city and more people moving to certain areas as they have the freedom to live where they would like to live rather than live close to their offices. Perhaps return-to-office initiatives also impact demand for home office supplies. All these factors can impact the printer supply stores and change quickly.

To explore this further, we've created a simulated dataset with sales from four stores serving a single metropolitan area that mimic the population movement trends heralded by COVID-19 policies and subsequent waffling on return-to-office versus remote work policies. Let's say Store 1 (*Figure 7.2*) serves a population that mainly works in nursing, grocery store management, and other jobs that do not stay remote for very long. This area also has temporary school closures. We'd expect to see a bump in sales initially during COVID-19 lockdowns but a quick return-to-normal pattern of sales:

Figure 7.2 – An illustration of a printer supply store – Store 1

However, Store 2 is located near the downtown area and serves a population of mostly mid-level technical workers, who largely don't fall under return-to-work policies once they are fully remote. Thus, sales spike during the initial lockdown and stabilize to the new normal for that area of town.

Store 3 serves a very desirable suburb near the downtown area mainly comprised of high-earning professionals who either worked from home or went into the office regularly before COVID-19. The commute is typically long, but the school system and safety of the neighborhood are excellent. During COVID-19, this suburb experienced a real estate boom as families elsewhere in the state decided to settle in that area now that they could work from home. Thus, by 2023, the area will have a large remote-worker population in need of printer supplies. Store 3 may need a larger space by 2023, as shown in *Figure 7.3*:

Figure 7.3 – An illustration of a warehouse of printer supplies

Store 4, however, serves the downtown area, which experiences a dramatic population loss as young families who like the convenience of a short commute move to the suburbs. Return-to-work policies don't attract workers back to condos in the downtown areas, leaving most buildings partially empty or inhabited by seasonal residents. Ideally, we'd like to avoid a pile-up of supplies and no potential clients (*Figure 7.4*):

Figure 7.4 – An illustration of an abandoned printer supply store with no customers

Our goal is to spot trends as they happen to pivot business strategy as quickly as possible to meet growing regional demands for supplies and to minimize real estate expenses as rents soar in our metro area. Given that these stores share a geography, we'll assume that the spatial weight matrix connects all four stores to each other. In a larger chain of stores, creating spatial weights based on demographic characteristics or distance from other stores might hone the analysis.

For now, let's turn our attention to the code that will help us find insights into our Burkina Faso market data and our COVID-19 store sales data.

Analyzing our spatiotemporal datasets

Our Burkina Faso market dataset is conveniently broken into quarters, and to understand quarter-by-quarter changes, it makes sense to set our window size as four quarters (a year) with a three-quarter overlap (to focus on monthly changes). As we did in *Chapter 2*, we'll calculate the local Moran statistic through the weight matrix and correlation metric. Let's set up our needed packages and import our Burkina Faso Market dataset plus its weight matrix with `Script 7.1`:

```
#import packages
import igraph as ig
from igraph import Graph
import numpy as np
import pandas as pd
import os
import matplotlib.pyplot as plt


#import Burkina Faso market millet prices
File="C:/users/njfar/OneDrive/Desktop/BF_Millet.csv"
pwd=os.getcwd()
os.chdir(os.path.dirname(File))
mydata=pd.read_csv(os.path.basename(File),encoding='latin1')


#import weight matrix of Burkina Faso markets
File="C:/users/njfar/OneDrive/Desktop/weights_bk.csv"
pwd=os.getcwd()
os.chdir(os.path.dirname(File))
weights=pd.read_csv(os.path.basename(File),encoding='latin1')
```

Now that we have our data and packages, let's calculate the Forman-Ricci curvature and PageRank centrality and save the network and overall metrics for each time slice by adding to `Script 7.1` with a correlation threshold set to `0.9` (for maximum sensitivity to major changes):

```
#score in yearly sets with 3 quarter overlap
vcurv_t=[]
vcurv_ave=[]
```

```
eig_t=[]
eig_ave=[]
nets=[]

for Year in range(4,29):
    data=mydata.iloc[(Year-4):(Year),1:46]
    cor=np.corrcoef(data.transpose())
    weights_total=weights.iloc[:,1:46]
    cor[cor>=0.9]=1
    cor[cor<0.9]=0
    cor_weighted=np.multiply(cor,weights_total)
    bf_market_w=Graph.Adjacency(cor_weighted,diag=False)
    edge_list=bf_market_w.get_edgelist()
    self_loop=[]

    for i in range(0,46):
        self=(i,i)
        self_loop.append(self)
    to_remove=[]

    for i in edge_list:
        for j in self_loop:
            if i==j:
                to_remove.append(i)

    bf_market_w.delete_edges(to_remove)
    nets.append(bf_market_w)
    d=Graph.degree(bf_market_w)
    e=Graph.pagerank(bf_market_w)
    eig_t.append(e)
    eig_ave.append(np.mean(e))
    ecurvw=[]

    for edge in bf_market_w.es:
        s=edge.source
        t=edge.target
        ecurvw.append(2-d[s]-d[t])
    vcurvw=[]

    for vertex in bf_market_w.vs:
        inc=Graph.incident(bf_market_w,vertex)
        inc_curv=[]
        for I in inc:
```

```
        inc_curv.append(ecurvw[i])
    vcurvw.append(sum(inc_curv))
vcurv_t.append(vcurvw)
vcurv_ave.append(np.mean(vcurv_t))
```

Now that we have our metrics calculated, let's plot the results to see how PageRank and Forman-Ricci curvature centrality vary over time by adding to `Script 7.1`:

```
#plot metric averages across time slices
time=range(0,25)
plt.plot(time, eig_ave, label = "PageRank Average")
plt.plot(time, vcurv_ave, label = "Forman-Ricci Curvature Average")
plt.xlabel("Time Slice")
plt.ylabel("PageRank and Forman-Ricci Curvature Averages")
plt.legend()
plt.show()
```

This should show a plot with very little variance in PageRank centrality but an increase and cyclical fluctuation of Forman-Ricci curvature centrality (*Figure 7.5*). This suggests that the volatility of millet prices increased over our markets, starting sometime around 2018-2019 (where volatility increased, then decreased, then increased again before leveling off at a higher volatility level) and got a bit worse during COVID-19 and the Ukraine conflict.

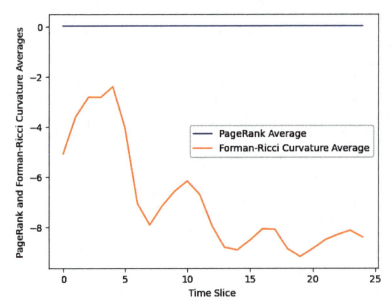

Figure 7.5 – A plot of average centrality metrics from 2015-2022 on the Burkina Faso market millet prices

Now that we have a plot showing changes in volatility, we can investigate what our markets look like at different periods. You may wish to investigate all periods, but we'll compare two periods—one with low volatility and one with high volatility—and show them in the following code. Let's plot time period 4, where volatility was low according to Forman-Ricci curvature centrality (highest centrality measurement in the dataset), and the size of each vertex is weighted by Forman-Ricci curvature for that vertex, by adding to Script 7.1:

```
#examine different time points with Forman-Ricci
#curvature plots, fourth slice
ig.plot(nets[4],vertex_size=np.array(vcurv_t[4])*-0.5)
```

This should yield a plot like *Figure 7.6*, showing very few markets connected by correlation and spatial weighting and very low Forman-Ricci curvature for vertices (very small vertex sizes):

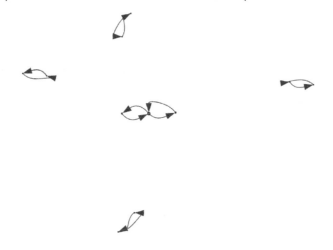

Figure 7.6 – A plot showing markets connected by correlation >0.9 within the
same geographical area weighted by Forman-Ricci curvature centrality

Figure 7.6 shows that most markets are not connected to each other, and those that are connected to each other tend to have very little mutual connectivity, yielding Forman-Ricci curvature centrality metrics near 0. This means that prices are independent, suggesting a robust network of markets that are unlikely to be highly impacted by external factors such as COVID-19 or international trade changes. Changes that impact one or two markets are unlikely to impact all of the country's markets, and sellers and consumers impacted by local issues (such as a dry quarter) can seek grain from a neighboring area if needed. The government can also address shortages by simply moving excess grain from one province with a good supply to another with supply issues. Fixing local issues with supplies won't take much time or money from a government perspective.

Let's examine a period of higher volatility (the 19th period) and see how connectivity and Forman-Ricci curvature centrality compared to time period 4 by adding to `Script 7.1`:

```
#examine different time points with Forman-Ricci
#curvature plots, 19th slice
ig.plot(nets[19],vertex_size=np.array(vcurv_t[19])*-0.5)
```

This should yield a plot like *Figure 7.7*, where many markets are connected to each other mutually, yielding Forman-Ricci curvature centrality metrics much further from zero, suggesting highly interconnected supply chains across large regions of the country that are vulnerable to large shortages that could cause a famine in the area. Addressing these issues will cost the government much more in terms of time and cost, leaving people vulnerable to food shortages or price spikes that make grain unaffordable for consumers.

> **Note**
>
> Some connected markets share many mutual connections, yielding strong curvature metrics, while some markets remain unconnected to any other market. The less-connected market prices suggest some areas are less vulnerable to widespread grain price spikes or shortages that would cause a famine.

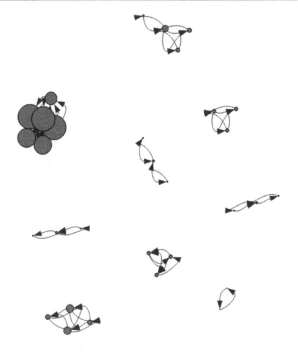

Figure 7.7 – The 19th time period in the Burkina Faso market millet price network

From our analyses, we can glean some insight into our market price behaviors over time. Prior to 2017, prices showed less volatility across markets. Individual markets weren't as tightly coupled with respect to price, suggesting local impacts on pricing. However, as markets became more dependent on global forces with respect to price, the markets became more volatile and susceptible to the impact of world events, such as COVID-19 and the Ukraine conflict. Indeed, in our raw dataset, we can see prices soaring after these events. This suggests that country-wide approaches are needed to address rising food prices, rather than focus on local markets that are struggling or vulnerable to changes in local production or global aid.

Looking at historical changes through the **United States Agency for International Development (USAID)** database and articles found on Google, we see that 2010-2020 was a period of development and international food aid in Burkina Faso, though many in the country relied mainly on subsistence farming rather than food purchased at markets. Given the rising prices, it is not surprising that subsistence farming and food aid were necessary, as grain prices reached crisis levels.

Now that we've seen how to analyze spatiotemporal data to visualize trends of volatility, let's turn our attention to the four stores impacted by work-from-home and return-to-office policies in a localized city. We'll ignore weighting, as these stores are in the same area and include only four stores. We'll use a threshold of 0.9 again to maximize our chances of finding extremely volatile periods in the data. Let's import our data with Script 7.2 and get started:

```
#import store sales data
File="C:/users/njfar/OneDrive/Desktop/Store_Sales.csv"
pwd=os.getcwd()
os.chdir(os.path.dirname(File))
mydata=pd.read_csv(os.path.basename(File),encoding='latin1')
```

We'll calculate Forman-Ricci curvature and PageRank centrality for this network by adding to Script 7.2; don't worry if you encounter a runtime warning, as it may appear for Windows machines but does not indicate a problem running your code:

```
#score in yearly sets with monthly  overlap
vcurv_t=[]
vcurv_ave=[]
eig_t=[]
eig_ave=[]
nets=[]


for Month in range(2,54):
    data=mydata.iloc[(Month-2):(Month),1:6]
    cor=np.corrcoef(data.transpose())
    cor[cor>=0.9]=1
    cor[cor<0.9]=0
    stores=Graph.Adjacency(cor)
```

```
edge_list=stores.get_edgelist()
self_loop=[]

for i in range(0,52):
    self=(i,i)
    self_loop.append(self)
to_remove=[]

for i in edge_list:
    for j in self_loop:
        if i==j:
            to_remove.append(i)
stores.delete_edges(to_remove)
nets.append(stores)
d=Graph.degree(stores)
e=Graph.pagerank(stores)
eig_t.append(e)
eig_ave.append(np.mean(e))
ecurvw=[]

for edge in stores.es:
    s=edge.source
    t=edge.target
    ecurvw.append(2-d[s]-d[t])
vcurvw=[]

for vertex in stores.vs:
    inc=Graph.incident(stores,vertex)
    inc_curv=[]
    for i in inc:
        inc_curv.append(ecurvw[i])
    vcurvw.append(sum(inc_curv))
vcurv_t.append(vcurvw)
vcurv_ave.append(np.mean(vcurv_t))
```

Again, we can plot PageRank and Forman-Ricci curvature centrality to identify periods of high and low volatility by adding to Script 7.2:

```
#plot metric averages across time slices
time=range(0,52)
plt.plot(time, eig_ave, label = "PageRank Average")
plt.plot(time, vcurv_ave, label = "Forman-Ricci Curvature Average")
plt.xlabel("Time Slice")
plt.ylabel("PageRank and Forman-Ricci Curvature Averages")
```

```
plt.legend()
plt.show()
```

You should see a plot similar to *Figure 7.8*, where volatility overall is fairly constant over the time period, with a few periods of increasing or decreasing volatility marking periods where pricing may be influenced by outside factors that our store managers cannot control.

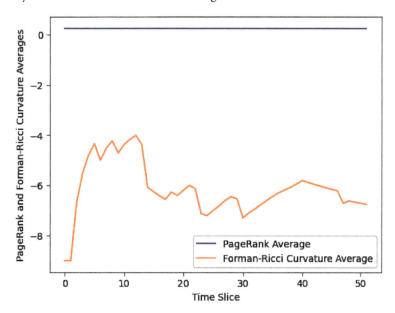

Figure 7.8 – A plot of PageRank and Forman-Ricci curvature averages

While we wouldn't expect major changes from period to period in volatility to show up in network plots for each period, we do see some differences in volatility that are worth investigating. Let's again choose two periods to investigate (period 10 and period 30). We'll investigate period 10 first by adding to `Script 7.2`, weighting our vertices by Forman-Ricci curvature to visualize the interconnectivity of sales:

```
#examine different time points with
#Forman-Ricci curvature plots, tenth slice
ig.plot(nets[10],vertex_size=np.array(vcurv_t[10])*-0.5)
```

This should show a plot like *Figure 7.9*, where only two stores are connected to each other, suggesting a period of relative stability where outside influences shouldn't impact sales across our stores as much as in other periods that are more volatile.

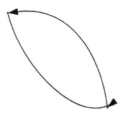

Figure 7.9 – A plot of our stores from period 10

Let's compare this plot with period 3 0's plot by adding to Script 7.2:

```
#examine different time points with
#Forman-Ricci curvature plots, thirtieth slice
ig.plot(nets[30],vertex_size=np.array(vcurv_t[30])*-0.5)
```

This code should produce a plot similar to *Figure 7.10*, showing connectivity across all four stores (maximum possible volatility), which leaves stores vulnerable to events that might impact sales across all four stores instead of impacting only one local store:

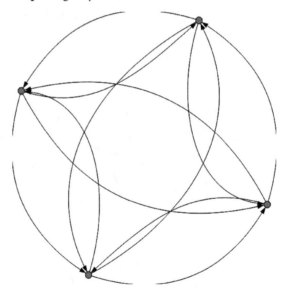

Figure 7.10 – A network plot of our store data from period 30 showing maximal volatility

Taken together, our results suggest volatility is fairly constant for our stores a few months prior to COVID-19 and increases upon the work-from-home mandates, continuing more or less through return-to-work policies. From our raw simulation data, some stores benefit from this change, while other stores suffer quite a bit. As volatility changes, store managers need to focus on factors they can control regarding their individual stores, and the manager overseeing all stores in the area may need to strategize on how to pivot supplies and growth needs of growing stores while considering the consolidation of buildings for those struggling stores.

Overall, these analyses provide some insight into pricing and sales vulnerability over times of crisis, allowing decision makers to pinpoint times requiring strategy pivots. Combined with other analytics (such as time series forecasting through multichannel singular spectrum analysis, for instance, or information on demographic or time-of-day differences in buying behavior), these types of analyses add insight into the timing and specific geographies that require attention to continue thriving in changing times.

Summary

In this chapter, we extended our prior insights on temporal and spatial data to tackle analytics problems with spatiotemporal datasets, including a store sales dataset and a Burkina Faso market millet price dataset. We sliced our two datasets by time, calculated spatial statistics, and analyzed network centrality metrics to identify changepoints and periods of volatility in our data. Periods of volatility correspond to network vulnerability to crashes and spikes in prices, sales, and other metrics, as the interconnectedness of the system allows for effects on one part of the network to spill into other parts of the network. We saw this at play during COVID-19 and the Ukraine conflict for Burkina Faso's markets, as well as the impacts of COVID-19 on the sales data. In *Chapter 7*, we'll look at sales and goods pricing across both time and geography to see how network science can solve problems in spatiotemporal data. In *Chapter 8*, we'll consider dynamic networks, where connections and even individuals can change over time, and examine how these changes impact disease spread in wildlife populations.

References

Han, X., Zhu, G., Zhao, L., Du, R., Wang, Y., Chen, Z., ... & He, S. (2023). *Ollivier–Ricci Curvature Based Spatio-Temporal Graph Neural Networks for Traffic Flow Forecasting. Symmetry, 15* (5), 995.

Schlör, J., Strnad, F. M., Fröhlich, C., & Goswami, B. (2022, May). *Identifying patterns of teleconnections, a curvature-based network analysis.* In *EGU General Assembly Conference Abstracts* (pp. EGU22-7256).

Shekhar, S., Jiang, Z., Ali, R. Y., Eftelioglu, E., Tang, X., Gunturi, V. M., & Zhou, X. (2015). *Spatiotemporal data mining: A computational perspective. ISPRS International Journal of Geo-Information, 4* (4), 2306-2338.

Wang, Y., Huang, Z., Yin, G., Li, H., Yang, L., Su, Y., ... & Shan, X. (2022). *Applying Ollivier-Ricci curvature to indicate the mismatch of travel demand and supply in urban transit network. International Journal of Applied Earth Observation and Geoinformation, 106,* 102666.

8

Dynamic Social Networks

In previous chapters, we've considered networks that don't change over time. Their geometry remains unchanged, and spreading processes and importance metrics calculated on them don't vary, either. However, this is not the case for many real-world networks, where new connections are forged and old connections break.

In this chapter, we'll consider networks that change over time and how these changes impact network structure and spreading processes. For instance, within a social network, users can join or deactivate accounts. The strength of their ties can change as they interact with other users' content. All these actions drive network changes over time, which impact how processes such as information exchange across a network function as hubs appear and disappear or bridges forge across different hubs. We'll consider simulation data and trial data collected from wildlife observations and hypothetical epidemics spreading across populations of crocodiles and blue herons, respectively.

By the end of this chapter, you'll understand how networks change over time and what these impacts mean for spreading processes and network metrics. You'll also know how to wrangle dynamic datasets into networks and analyze time slices. These methods scale to larger populations and longer time frames, so you'll be equipped to work with population models at scale by the end of this chapter.

In this chapter, we will cover the following topics:

- Social networks that change over time
- A deeper dive into spreading on networks
- Example with evolving wildlife interaction datasets

Technical requirements

In this chapter, we will be using a Jupyter notebook to run our dynamic network examples. You can also run this code on a cloud platform that supports Jupyter notebooks, such as **Amazon Web Services** (**AWS**) or **Google Cloud Platform** (**GCP**). We will be using a Jupyter notebook run on a laptop computer, and it is not necessary to have cloud access to process the examples in this chapter.

Social networks that change over time

Consider your group of closest friends. How has that group changed over the last year? The last 5 years? The last 10?

Many social relationships and other types of data that can be represented using networks evolve over time. Friendships drift away and forge as people move, graduate from school, get married, change jobs, travel, or pass away. Animals migrate with changes in season, changes in climate, and human activity encroaching on grazing lands. Sales patterns across stores can be interrupted by new stores in the area, movements of population, and major world events such as COVID-19 or the Ukraine conflict.

In the next sections, we'll explore dynamic networks, whose structure changes over time. As network structure changes, so do many important properties related to information/disease flow and partitioning of the network.

Friendship networks

Many social network platforms allow friendship groups to evolve and even encourage changes in structure. Consider sharing posts or tweets, tagging other people in your network post/tweet/picture who might have friends who have not met you yet, and adding searchable hashtags related to the content you are sharing. Each of these activities provides more visibility for you outside your initial social network, allowing friends of friends, acquaintances who haven't connected with you on the social media platform, and individuals who share similar interests in your topic post to find you and connect.

Let's say someone posted a surreal dreamcatcher image on a social media platform (*Figure 8.1*):

Figure 8.1 – A dreamcatcher image created with artificial intelligence (AI) posted to a social media platform

Perhaps the user added #AIArt and #Dreamcatcher and tagged a friend who posts similar content to *Figure 8.1*. Other platform users may search for others creating images with AI art platforms and want to follow the user or like/share the content. This provides more connections for the user and expands their network.

The friend who is tagged may benefit from these new connections, as the new connections for this user visit the friend's page. Additionally, friends of that friend who haven't encountered the user may also connect with the user or like/share their content. This also expands the user's network.

Many social media platforms also connect users by creating interest pages or groups where users with similar interests can post content, connect, and discuss the topic. For instance, many social media platforms provide animal interest groups or pages with cute animal art or photographs, such as the animal art shown in *Figure 8.2*:

Figure 8.2 – An AI art image of a baby elephant in a teacup surrounded by roses

These interest groups encourage connections between users and the sharing of searchable content to grow the group's membership (sometimes encouraged by advertising optimization algorithms or other paid users, including groups themselves in some cases). Popular content tends to encourage group growth and interactions/connections between users in the group. Unpopular content tends to discourage growth.

The darker side of algorithms that encourage connection generation between users with similar interests and optimize content spread is the potential for conspiracy theories and misinformation spread, as well as promoting connections between racist or discriminatory groups. During the 2016 and 2020 United States presidential elections (and during the COVID-19 pandemic), misinformation spread rapidly across social media platforms. During pandemics or national crises, misinformation can create public panic, increase strain on resources, or precipitate higher death rates. Within the political realm, misinformation can sway election results or accelerate divides between different political groups.

Algorithms that encourage the evolution of a social network to increase the number of likes by connecting people who probabilistically will like other users' content or views of related advertisements can inadvertently do the following:

- Promote the spread of misinformation among users likely to like or comment on it

- Encourage connections between users sharing misinformation or harmful content

- Create a feedback loop that perpetuates this problem on the social network

This can happen on social media platforms as well as community groups, internet chatrooms, and other forums. While promoting more cat memes may not be harmful (minus the potential for time wasted looking at cat memes), promoting racist ideologies or encouraging religious violence is quite harmful.

Triadic closure

Most of the network evolution algorithm impacts result from a network property called **triadic closure**. Triadic closure on social networks roughly relies on the premise that friends of a user's friends in a network might know the user in real life but not be connected yet. This is a direct connection-based instance of triadic closure in social networks. Another user who shares a friend group with many of our user's friends likely knows that user (or would be a good connection for the user based on the shared social network). In addition, other users might share many interests with the user (the premise of many friend suggestion algorithms); this is a metadata-based instance of triadic closure, where connection formation hinges on similarity of user metadata rather than shared connections.

Consider Pieter, Kwame, and Pablo. All three enjoy football, software engineering, and hiking. Pablo and Pieter know each other, as do Pieter and Kwame; however, we don't know if Kwame and Pablo know each other (*Figure 8.3*):

Kwame

Pieter

Pablo

Figure 8.3 – A diagram of the known relationships among our three football fans

Given the two relationships that are known, along with the shared interest metadata, it is likely that Kwame and Pablo know each other, as well (or would get along if they connected). The percentage of open triads on a network measures the overall triadic openness of the network. Networks with a high percentage of open triads have a high potential for evolution over time as friends of friends or individuals with similar interests connect over shared content or group membership. Now that we know a bit more about relationships and how they can evolve over time, let's dig deeper into spreading processes over evolving networks.

A deeper dive into spreading on networks

Now that we understand a bit about how networks can change over time, let's see how we can represent these networks in Python and compute spreading processes such as epidemic models over time on a dynamic network. Python provides several tools to help us both visualize and analyze dynamic networks as we have in prior chapters. Let's get started.

Dynamic network introduction

We can track changes over a network through time by keeping track of connections at different time points. Let's consider a pride of lions in which individuals interact periodically over the course of the day. *Figure 8.4* shows this example network with timestamps to track edges:

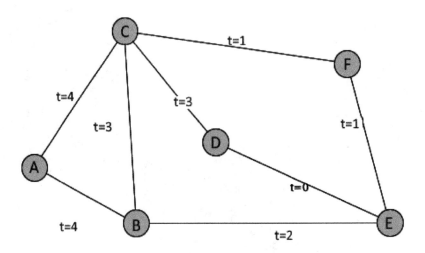

Figure 8.4 – An example of a dynamic network with timestamps

Figure 8.4 shows an example of a dynamic lion network with an evolving structure of edges and vertices that can be stamped with time information. Python provides a handy package to create a network such as this. The `NDlib` package utilizes `DyNetx` to model time-evolving graphs, which utilizes several key primitives from `DyNetx` to create and analyze dynamic networks. This provides a powerful framework for working with dynamic networks in the real world. This is particularly valuable in fields such as epidemiology, social network analysis, and transportation systems, where understanding temporal dynamics is essential. Let's explore this package by creating *Figure 8.4* using `Script 8.1`:

```
#create Figure 8.4
#import packages
#install dynetx if needed
#!pip install dynetx
import dynetx as dn
import networkx as nx
import matplotlib.pyplot as plt
#create an empty graph
g = dn.DynGraph()
#add vertices and edges with time information added

#adding the edge ("D","E") at t=0 that vanishes at time e=2
g.add_interaction(u="D", v="E", t=0, e=2)

# adding some edges at time t=1
g.add_interactions_from([[("C", "F"), ("F", "E")], t=1)
```

```
# adding some edges at time t=2
g.add_interactions_from([("B", "E"), ("A", "B")], t=2)

# adding some edges at time t=3
g.add_interactions_from([("B", "C"),("C","D")], t=3)

# adding some edges at time t=4
g.add_interaction(u="A", v="C", t=4)
```

Once we have *Figure 8.4* created, we can return a list of snapshots, each of which represents a static view of the lion network at different points in time. These snapshots are essentially instances of the graph at different timestamps. Let's add to Script 8.1 to show these plots:

```
#create plots for each time point
for i in range (5):
    g1 = g.time_slice(i)
    nx.draw(g1,with_labels=True)
    print ("snapshot at t = ", i)
    plt.show()
```

We can print the list of edges for each snapshot, as well. Let's add to Script 8.1 to get a list of edges at each time point:

```
#create lists of edges at each time point
for i in range (5):
    g1 = g.time_slice(i)
    print ("snapshot at t = ", i)
    print(g1.edges())
```

This should give us the following analysis of edges at each time point, shown next:

```
snapshot at t =  0
[('D', 'E')]
snapshot at t =  1
[('D', 'E'), ('E', 'F'), ('F', 'C')]
snapshot at t =  2
[('E', 'B'), ('B', 'A')]
snapshot at t =  3
[('D', 'C'), ('C', 'B')]
snapshot at t =  4
[('C', 'A')]
```

Understanding edge structure at each time point is critical in the assessment of network structure and properties that facilitate flow on the network, such as hubs and bridges. Visualization and edge lists provide us a way to summarize information, as well, which might provide added insight into, say, an epidemic through a small population. Moving away from our lion example, let's consider badge tracking in an office setting during the outbreak of a new disease (such as COVID-19). Different individuals will interact with different people and different numbers of people in the office on different days, depending on their schedules. Lists of edges that exist at different time points and visualizations of interactions among employees can help researchers track the new outbreak and calculate important epidemic model parameters to understand the potential impact of the disease on a larger population.

Just as connections between individuals can change over time, so can the individuals in the network themselves. Perhaps one employee is out sick or takes a personal day, thus dropping out of the network at that time point. With DyNetx, we can also create an interaction network to model a dynamic structure in which both vertices and edges can appear and disappear as time evolves. Returning to *Figure 8.4*, let's add to Script 8.1 to create the following interaction stream:

```
#print interaction stream
for i in g.stream_interactions():
    print(i)
```

This script creates an interaction list for each time point, denoting interactions that exist in the network. While we don't have individual lions leaving or entering our network, lion interactions do change, and lions may or may not have interactions with other lions in the network for a given time point. Let's look a bit further at the output to see which time points involve the most and least interaction among the network lions:

```
#output of interactions for each time point
('D', 'E', '+', 0)
('C', 'F', '+', 1)
('F', 'E', '+', 1)
('D', 'E', '-', 2)
('B', 'E', '+', 2)
('A', 'B', '+', 2)
('B', 'C', '+', 3)
('C', 'D', '+', 3)
('A', 'C', '+', 4)
```

The first two values specify the vertices participating in an edge. The third value designates the type of edge operation, with '+' indicating an appearance (edge creation) and '-' denoting vanishing (edge removal). The last value signifies the timestamp associated with the edge operation. For instance, we can see that lion F establishes multiple interactions within the network at time point 1. We can also see that lions D and E, whose interaction began at the 0th time point, cease interaction at time point 2.

This type of information is critical for epidemic tracking, as it allows us to narrow down contacts to perform contact tracing. **Contact tracing** involves working backward through time and interactions among impacted individuals to find the original patient (**patient zero**) who was infected with the disease of interest. This allows epidemiologists to determine the geographic region from which the outbreak started as well as trace its evolution over time and space from patient zero. Some diseases mutate rapidly, which complicates the development of a treatment. What works today may not work with the next mutation. This is why flu vaccines are typically given annually, as flu strains mutate regularly. However, some diseases mutate very slowly or not at all, allowing for effective treatments or vaccines.

Now that we know a bit about dynamic network construction in Python, let's explore **susceptible-infected-resistant** (**SIR**) models in a bit more depth than we did in *Chapter 3*.

SIR models, Part Two

In *Chapter 3*, we explored SIR models to track network spreading processes, but we did not go into technical detail regarding the differential equations. In this section, we will dive a bit further into SIR model mathematics. Let's start by defining the terms we'll use in our model:

- $S(t)$ represents the number of susceptible individuals at time t: those who do not have the disease but can contract it

- $I(t)$ represents the number of infectious individuals at time t: those who are actively infected and able to pass along the disease to susceptible individuals

- $R(t)$ represents the number of recovered individuals at time t: those who no longer are infected and cannot be reinfected due to immunity

- beta (β) is the transmission rate of infected individuals to a population of susceptible individuals

- gamma (γ) is the recovery rate (the rate at which infected individuals become recovered individuals)

- N is the total population size

The populations of susceptible, infected, and recovered individuals will change over time as individuals become infected and later recover. We can define the population numbers at each time point by relating these populations to the transmission and recovery rates in a series of equations defined by time. The susceptible population will decline as disease is transmitted from infected individuals to susceptible individuals. The infected population may grow or decline at a given time point as more susceptible individuals become infected and as more infected individuals recover. The recovered population will grow over time as infected individuals recover from the disease.

For more technically minded readers, we can arrange these terms to define a system of differential equations governing spread and recovery within a population of susceptible individuals, as shown next:

$$\frac{dS}{dt} = -\frac{\beta IS}{N},$$
$$\frac{dI}{dt} = \frac{\beta IS}{N} - \gamma I,$$
$$\frac{dR}{dt} = \gamma I$$

A high transmission rate coupled with a low recovery rate will yield a large population of infected individuals. A low transmission rate with a high recovery rate will result in a slow or fizzling-out epidemic in the population.

However, interactions within a population are rarely random, as assumed by the SIR equations. Dynamic networks provide a way to capture interactions over time, leading to more accurate models of transmission within a population of interest or allowing for more accurate planning through simulated models of interaction levels within a population of interest. Let's build on *Chapter 3*'s discussion of epidemics by introducing some new network metrics that influence epidemic spread.

Factors influencing spread

In *Chapter 3*, we examined the importance of bridges and hubs to the spread of a music trend or epidemic across networks. Since then, we've considered many network centrality metrics that relate to the concepts of hubs and bridges. Let's briefly review a few of these centrality metrics.

Hubness can be assessed with several centrality metrics, including the following:

- Degree centrality (defined in *Chapter 6*)
- PageRank centrality (defined in *Chapter 6*)
- Hubness centrality (defined in *Chapter 6*)

Degree centrality is the simplest and most related to spreading processes on a network. The number of connections an infected vertex has to other vertices impacts the ability of an infected vertex to spread the disease to its immediate neighbors. The combination of degree centralities across the network impacts the overall ability of a disease to spread globally. Sparse networks with very low degree centralities don't offer chances for the disease to infect many individuals at the same time, leading to a situation where public health workers can effectively implement strategies such as isolation and contact tracing without being overwhelmed by many infected patients at a given time point.

However, networks with very high degree centralities likely include many hubs or a single large hub. A large hub allows one infected individual to pass along the disease to many others, who also have many contacts who can become infected. This means that public health resources are easily overwhelmed at the start of an epidemic, and the chances of containing such an epidemic before many people are infected are quite low. We saw this in many global cities during COVID-19, and citywide shutdowns became necessary to avoid overwhelming healthcare systems; rural areas did not experience the same rapid spikes of infection rates that we saw in large cities.

Bridges also play an important role in epidemic modeling on networks. If two large hubs are not connected, it's easy to isolate one hub from the other to stop an epidemic from spreading from one hub to the other, as we simply ensure that no bridges are built between the two hubs until the epidemic has ended in the infected hub. Bridges connecting the two hubs provide an opportunity for the bridge vertex/vertices to become infected and pass the infection from one community to another. Thus, bridges provide an opportunity for infection to spread and become a regional or global problem rather than a local community problem.

Recall that bridging properties can be measured through **betweenness centrality** (defined in *Chapter 6*), which measures the number of a network's shortest paths that pass through a vertex. Essentially, this measures the importance of each vertex to spreading processes on the network. A vertex with a high betweenness centrality represents a high chance of passing a disease to other parts of a population should that vertex become infected. Isolating or vaccinating vertices with a high betweenness centrality can disrupt disease spread on a network, and these individuals represent a chance to make a big impact on overall epidemic severity with limited resources. Within the context of COVID-19, frequent international travelers served as bridges between countries. Halting international travel from infected areas to areas not yet impacted allowed small nations, such as those on Caribbean islands, to avoid epidemic spread to their populations.

Forman-Ricci curvature centrality (defined in *Chapter 6*) accounts for both bridging properties and hub properties within a network by considering edges connected to other edges through a set of vertices. Recent studies suggest that Forman-Ricci curvature can find edges and vertices important to spreading processes and, thus, serves as an effective metric to use in epidemic simulations where vertices are isolated or otherwise taken out of a population (through vaccination, for instance).

Eccentricity, the maximum distance from a vertex to another vertex in a graph, is another metric commonly used to measure the potential impact of an epidemic at a population level. The **radius** of a network is the smallest eccentricity value across network vertices, and the **diameter** of a network is the largest eccentricity value across network vertices. Dense networks have smaller diameters than sparse networks, as vertices tend to be more interconnected (allowing for greater spread potential in shorter periods of time).

Both radius and diameter bound the behavior of differential equations, such as our SIR model, on a network. While the theoretical proofs are beyond the scope of this book, we'll explore radius and diameter in practice with a real-world simulation of two wildlife epidemic datasets in the next section. Let's get started by exploring our datasets.

Example with evolving wildlife interaction datasets

We collected two datasets on within-species wildlife interactions (interactions involving animals of the same species) over short periods of time.

The first dataset includes the observation of 4 crocodiles in a South Florida lagoon across 15-minute intervals over 12 time periods after a cold front passed through the area, prompting more crocodiles to seek out sunny shorelines. Social interactions between crocodiles were defined as animals that interacted within a single 15-minute period. Connections were noted for each 15 minutes, giving a dynamic network of crocodile social interactions across 12 time periods.

The second dataset includes the observation of 7 blue herons interacting at a bait ball of migrating fish in a South Florida lagoon captured across 5-minute intervals over 10 time periods. Interaction among blue herons was defined as herons occupying the same cubic meter of space above the bait ball. Connections were noted for each 5 minutes, giving a dynamic network of blue heron social interactions across 10 time periods. Populations were initially separated into two nesting communities, with two herons in one community and five herons in the other community.

Let's get started by examining the epidemic dynamics and centrality metrics on the Crocodile Network dataset.

Crocodile network

We'll first consider our crocodile network, where 4 crocodiles interact over the course of 3 hours. Script 8.2 loads the relevant packages and creates our initial set of networks over time:

```
#import packages
import networkx as nx
import dynetx as dn
from past.builtins import xrange
import matplotlib.pyplot as plt

#create a dynamic graph object
crocodile = dn.DynGraph() # empty dynamic graph

#fill in relevant interactions
crocodile.add_interaction(u="Croc1", v="Croc2",t=0)
crocodile.add_interactions_from(
    [("Croc1","Croc2"),("Croc3","Croc4")],t=1)
crocodile.add_interactions_from(
    [("Croc1","Croc2"),("Croc3","Croc4")],t=2)
crocodile.add_interactions_from([[("Croc1","Croc2")],t=3)
crocodile.add_interactions_from(
    [("Croc1","Croc2"),("Croc2","Croc3")],t=4)
```

```
crocodile.add_interactions_from(
    [("Croc1","Croc2"),("Croc2","Croc3")],t=5)
crocodile.add_interactions_from([[("Croc1","Croc2")],t=6)
crocodile.add_interactions_from([[("Croc3","Croc4")],t=7)
crocodile.add_interactions_from([[("Croc3","Croc4")],t=8)
crocodile.add_interactions_from(
    [("Croc2","Croc3"),("Croc3","Croc4")],t=9)
crocodile.add_interactions_from([[("Croc2","Croc3")],t=10)
crocodile.add_interactions_from(
    [("Croc1","Croc2"),("Croc2","Croc3")],t=11)

#plot time slices
for i in range(12):
    g1 = crocodile.time_slice(i)
    nx.draw(g1,with_labels=True)
    plt.show()
```

Figure 8.5 shows the second time point plot, where two sets of crocodiles interacted with another crocodile:

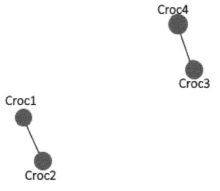

Figure 8.5 – The second time period plot of the crocodile network

We can track the changes in crocodile interaction over time by adding to Script 8.2:

```
for i in crocodile.stream_interactions():
    print(i)
```

This gives the following interaction changes at time points where they occurred:

```
('Croc1', 'Croc2', '+', 0)
('Croc3', 'Croc4', '+', 1)
('Croc2', 'Croc3', '+', 4)
('Croc1', 'Croc2', '-', 7)
('Croc3', 'Croc4', '+', 7)
```

```
('Croc2', 'Croc3', '+', 9)
('Croc3', 'Croc4', '-', 10)
('Croc1', 'Croc2', '+', 11)
('Croc2', 'Croc3', '-', 12)
```

We can see that interactions change slowly and then become more frequent toward the end of our observation period. This may impact the dynamics of an epidemic, as crocodiles have more varied interactions later in the observation period. In addition, early additions of interactions provide the opportunity for a disease to spread across the population, as crocodiles begin to interact with each other.

Let's consider an epidemic that starts with the infection of one crocodile with a high infection rate (beta = 0.3) relative to the recovery rate (gamma = 0.05). This corresponds to a moderately infectious disease passing through our crocodile population that results in a long infection period. Crocodile pox is one such disease that can cause mortality in juvenile crocodiles. Infected animals tend to be symptomatic for weeks or months. Infection of cells tends to happen rapidly with a low incubation period (minutes). While the exact parameters for infection rate and recovery rate are not known, these estimates are good proxies for the infection. Let's add to Script 8.2 to simulate crocodile pox spread in our population of crocodiles as they interact over 12 time periods:

```
#run epidemic model
import ndlib.models.ModelConfig as mc
%matplotlib inline
from ndlib.viz.mpl.DiffusionTrend import DiffusionTrend
import ndlib.models.dynamic as dm


#model selection
model = dm.DynSIRModel(crocodile)

#model configuration
config = mc.Configuration()
config.add_model_parameter('beta', 0.3) # infection rate
config.add_model_parameter('gamma', 0.05) # recovery rate
config.add_model_parameter("percentage_infected", 0.25)
model.set_initial_status(config)

# simulate snapshot based execution
iterations = model.execute_snapshots()

#iterations = model.execute_iterations()
trends = model.build_trends(iterations)

#visualize trends
viz = DiffusionTrend(model, trends)
viz.plot()
```

Figure 8.6 summarizes the dynamics of this epidemic among our crocodile sample on our run of the algorithm. Given that this is a probabilistic model, your epidemic may proceed a bit differently with respect to time points or the number of infected crocodiles at a given point in time. The plot summarizes the number of susceptible, infected, and removed (recovered) crocodiles at each time point of the epidemic.

In our trial, this crocodile pox epidemic starts rapidly with the infection of a single crocodile at period 0 and quickly infects the entire population by period 5, with none of the animals recovered by the end of observation. Any new animals introduced into the population who were not immune to crocodile pox would likely catch crocodile pox soon after entering the population and remain sick throughout the observation period. This suggests a need to quarantine the infected crocodiles to avoid epidemic spread in wild populations (perhaps by isolation and treatment in a local rehabilitation center):

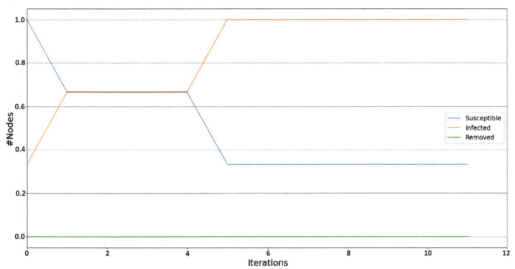

Figure 8.6 – The viral dynamics of crocodile pox in our crocodile network

We can examine how key network metrics differ over time to examine any changes that might have contributed to the spread of crocodile pox in our crocodile network animals. From our construction of the dynamic network, we can see increases in population mixing from periods 0-5 in the Crocodile Network dataset, where all crocodiles can infect each other by the end of period 5 (corresponding to the time point where all crocodiles, in fact, have been infected). Let's add to Script 8.2 so that we can calculate betweenness centrality across the vertices in the network and the overall radius and diameter of the network at each time point:

```
#obtain network centrality statistics for crocodile network
for i in range(12):
    dg = crocodile.time_slice(i)
    try:
        # computing betweenness centrality
```

```
        dd = nx.betweenness_centrality(dg)
        plt.plot(dd.values())
        plt.title("betweenness centrality")
        print()
        plt.show()
        print ("timestamp = ", i,",", "nodes =",
            len(dg.nodes()),",",
            "max degree = ",max(dg.degree().values()))
        print ("diameter = ", nx.diameter(dg))
        print ("radius = ", nx.radius(dg))
    except:
        print('Infinite value observed.')
```

Note that we have infinite values observed when vertices are isolated. We'll skip calculations for those time periods by adding a try/except clause to our script. The output should show a plot like *Figure 8.7* (corresponding to the first time point), showing betweenness centrality across vertices:

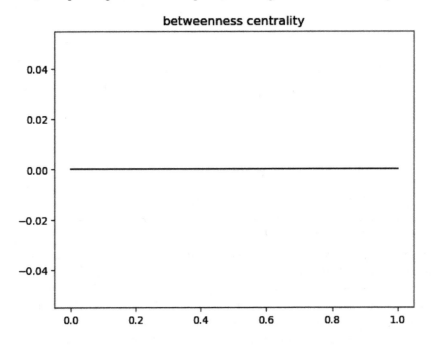

Figure 8.7 – Betweenness centrality plot for the first time point of observation in the crocodile network

Below each plot, we can see a summary of the timestamp, the number of connected vertices, the maximum degree of the network, the network radius, and the network diameter for time points where connected vertices exist. *Figure 8.8* shows the output from the first time point of observation, where the radius and diameter of the network are both 1:

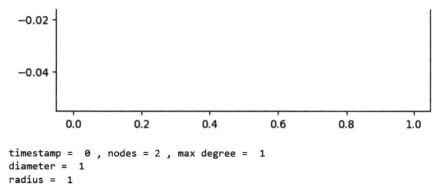

```
timestamp =  0 , nodes = 2 , max degree =  1
diameter =  1
radius =  1
```

Figure 8.8 – Output for the first time point of observation in the Crocodile Network dataset

Note that the maximum diameter observed in our dynamic Crocodile Network dataset is a diameter of 2. Several of these larger diameters occur during the infection period, where more crocodiles are interacting with each other (and forming new interactions). In real-world epidemics, population mixing plays an important role in epidemic spread dynamics, with more mixing tending to produce more severe epidemics that include more cases across populations and a longer duration of epidemic.

Now that we understand a bit more about dynamic networks and epidemic models, let's turn to a larger Heron Network dataset, where we have individuals from two separate nesting populations interacting at a bait ball, which is closer to how a real epidemic involving multiple populations might start.

Heron network

Let's get started with our Heron Network dataset. Script 8.3 creates a heron network and plots each time point in this network:

```
#create empty graph
blue_heron = dn.DynGraph() # empty dynamic graph

#add relevant interactions
blue_heron.add_interactions_from(
    [("h1","h2"),("h2","h3"),("h3","h4"),("h4","h5")],t=0)
blue_heron.add_interactions_from(
    [("h1","h2"),("h2","h3"),("h3","h4"),("h5","h6")],t=1)
blue_heron.add_interactions_from(
    [("h1","h2"),("h2","h3"),("h5","h6"),("h6","h7")],t=2)
```

```
blue_heron.add_interactions_from(
    [("h1","h2"),("h3","h4"),("h6","h7")],t=3)
blue_heron.add_interactions_from(
    [("h2","h3"),("h3","h4"),("h4","h5"),("h6","h7")],t=4)
blue_heron.add_interactions_from([[("h3","h4")],t=5)
blue_heron.add_interactions_from(
    [("h1","h2"),("h2","h3"),("h3","h4"),("h4","h5"),
        ("h5","h6")],t=6)
blue_heron.add_interactions_from(
    [("h1","h2"),("h2","h3"),("h3","h4"),("h4","h5"),
        ("h5","h6"),("h6","h7")], t=7)
blue_heron.add_interactions_from(
    [("h1","h2"),("h2","h3"),("h3","h4"),("h5","h6"),
        ("h6","h7")],t=8)
blue_heron.add_interactions_from(
    [("h1","h2"),("h3","h4"),("h5","h6"),("h6","h7")],t=9)

#plot network over time
for i in range(10):
    g1 = blue_heron.time_slice(i)
    nx.draw(g1,with_labels=True)
    print()
    plt.show()
```

Figure 8.9 shows the first time point in our heron network, where the five herons from a nesting community have interacted with each other at our bait ball:

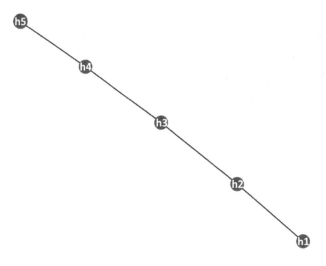

Figure 8.9 – The first time point plot of our heron network

Note that our heron network shows many different interaction patterns over the observation duration. This will impact our viral dynamics, as different animals interact at different time periods. Let's consider an infectious disease that spreads rapidly (beta = 0.4) and resolves at a slower pace (gamma = 0.2). Bird flu epidemics tend to spread rapidly, and these parameters may represent a novel bird flu variant passing through a population. In a real epidemic of bird flu, longer latency to infection means we'd need to observe the population over a period of longer than 50 minutes (perhaps over the course of several days). However, let's see how our proposed epidemic spreads through our heron population based on interactions by adding to Script 8.3:

```
#run epidemic model
import ndlib.models.ModelConfig as mc
%matplotlib inline
from ndlib.viz.mpl.DiffusionTrend import DiffusionTrend
#from ndlib.viz.mpl.DiffusionPrevalence import DiffusionPrevalence
import ndlib.models.dynamic as dm

# model selection
model = dm.DynSIRModel(blue_heron)

# model configuration
config = mc.Configuration()
config.add_model_parameter('beta', 0.4) # infection rate
config.add_model_parameter('gamma', 0.2) # recovery rate
config.add_model_parameter("percentage_infected", 0.1)
model.set_initial_status(config)

# simulate snapshot based execution
iterations = model.execute_snapshots()
#iterations = model.execute_iterations()
trends = model.build_trends(iterations)

#visualize trends
viz = DiffusionTrend(model, trends)
viz.plot()
```

This simulation shows an epidemic that infects nearly half of our heron population by the end of our observation period, with low levels of infection and recovery by the second time point. New infections occur periodically, and one heron appears to have a longer infection period than expected (period 4–7). *Figure 8.10* visualizes the dynamics of this epidemic (which may differ in your run, given that the infection model is probabilistic):

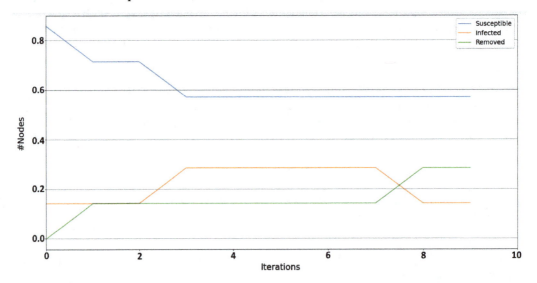

Figure 8.10 – A visualization of an epidemic on our heron network

Were this epidemic to happen in a real population of herons, it may be possible to quarantine the infected heron showing a long infection period to avoid further spread in the population. In conserved populations such as those in wildlife preserves or rehabilitation centers and populations bred for human consumption, this strategy is common to avoid disease spread across animals. In the wild, it is difficult to identify infected animals or even detect an outbreak until many animals turn up sick in areas that are monitored.

Let's now consider network metrics over the course of the epidemic by adding to Script 8.3, again adding a try/except clause, as herons are isolated at times during the period of observation:

```
#obtain network centrality statistics for the blue heron network
for i in range(10):
    dg = blue_heron.time_slice(i)
    try:
        # computing betweenness centrality
        dd = nx.betweenness_centrality(dg)
        plt.plot(dd.values())
        plt.title("betweenness centrality")
        print()
```

```
        plt.show()
        print ("timestamp = ", i,",", "nodes =",
            len(dg.nodes()),",",
            "max degree = ",max(dg.degree().values()))
        print ("diameter = ", nx.diameter(dg))
        print ("radius = ", nx.radius(dg))
    except:
        print('Infinite value observed.')
```

For the first period of observation, we obtain a betweenness centrality graph that shows large variations in values, as shown in *Figure 8.11*:

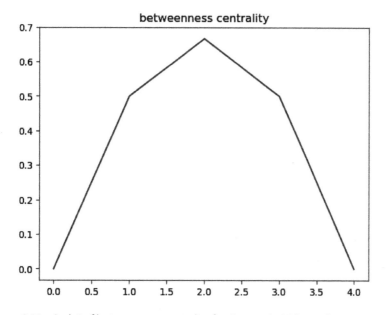

Figure 8.11 – A plot of betweenness centrality for time period 1 in our heron network

Betweenness centrality varies greatly across the period of observation, from nearly 1 to 0. The diameter varies from 1 (time point 6) to 6 (time point 8); the radius varies from 1 (time point 6) to 3 (time points 7 and 8). Isolated vertices are common, and we see the population split up in interactions (time points 3 and 9), where some herons interact in different locations and some do not interact at all. *Figure 8.12* shows a split heron population example:

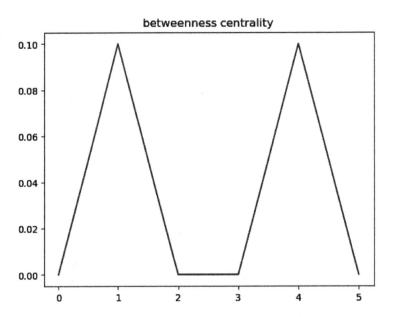

Figure 8.12 – A split heron population interaction pattern, observed at point 3

Frequent changes in interaction patterns likely limited this epidemic's potential spread, as animals interacted infrequently enough and for short enough time periods to limit exposure to infectious individuals relative to the frequent and lasting interactions noted in the crocodile population. In the wild, the frequency of interaction and duration of interaction among animals of the same species play a critical role in determining the likelihood of population epidemics within a species sharing a common geography. Solitary species tend to have lower rates of infectious disease spread than species forming large herds or flocks.

Naturally occurring phenomena such as bait balls or temporary water sources after a storm provide time-limited interaction opportunities for disease to spread among solitary species or isolated herds. As such, they present an opportunity for conservationists to study epidemic potentials within populations of animals that may not interact frequently aside from a one-time event involving needed resources such as food or water. Models such as the ones we've examined provide the tools for estimating the impact of infectious disease on populations that are normally isolated but mix at such events.

Summary

In this chapter, we explored dynamic networks and how changes in network structure over time impact epidemic spread using examples of disease spread among crocodile and blue heron populations. We also explored the relationship between network metrics and epidemic spread, noting patterns of connectivity associated with higher rates of spread and more severe epidemics within a hypothesized population. These tools are critical to understanding trends and epidemics spread across real-world networks, allowing researchers to plan out critical infrastructure to deal with crises such as COVID-19 or Ebola in human populations, infectious diseases threatening endangered animal populations, fake news before a country's elections, or dangerous behavioral trends spreading among youth on social media.

In the next chapter, we will shift focus to **machine learning (ML)** on networks.

References

Bianconi, G., Darst, R. K., Iacovacci, J., & Fortunato, S. (2014). Triadic closure as a basic generating mechanism of communities in complex networks. *Physical Review E, 90(4), 042806.*

Büttner, K., Krieter, J., Traulsen, A., & Traulsen, I. (2016). Epidemic spreading in an animal trade network–Comparison of distance-based and network-based control measures. *Transboundary and Emerging Diseases, 63(1), e122-e134.*

Fèvre, E. M., Bronsvoort, B. M. D. C., Hamilton, K. A., & Cleaveland, S. (2006). Animal movements and the spread of infectious diseases. *Trends in microbiology, 14(3), 125-131.*

Ganesh, A., Massoulié, L., & Towsley, D. (2005, March). The effect of network topology on the spread of epidemics. *In Proceedings IEEE 24th Annual Joint Conference of the IEEE Computer and Communications Societies. (Vol. 2, pp. 1455-1466). IEEE.*

Garin, F., Varagnolo, D., & Johansson, K. H. (2012). Distributed estimation of diameter, radius and eccentricities in anonymous networks. *IFAC Proceedings Volumes, 45(26), 13-18.*

Giuggioli, L., Pérez-Becker, S., & Sanders, D. P. (2013). Encounter times in overlapping domains: application to epidemic spread in a population of territorial animals. *Physical review letters, 110(5), 058103.*

Liu, C., & Zhang, Z. K. (2014). Information spreading on dynamic social networks. *Communications in Nonlinear Science and Numerical Simulation, 19(4), 896-904.*

Lou, T., Tang, J., Hopcroft, J., Fang, Z., & Ding, X. (2013). Learning to predict reciprocity and triadic closure in social networks. *ACM Transactions on Knowledge Discovery from Data (TKDD), 7(2), 1-25.*

Martcheva, M. (2014). Avian flu: modeling and implications for control. *Journal of Biological Systems, 22(01), 151-175.*

Milli, L. (2018). Understanding spreading and evolution in complex networks.

Moore, R. L., Isberg, S. R., Shilton, C. M., & Milic, N. L. (2017). Impact of poxvirus lesions on saltwater crocodile (Crocodylus porosus) skins. *Veterinary microbiology, 211, 29-35.*

Read, J. M., Eames, K. T., & Edmunds, W. J. (2008). Dynamic social networks and the implications for the spread of infectious disease. *Journal of the Royal Society Interface, 5(26), 1001-1007.*

Sarker S, Isberg SR, Moran JL, Araujo R, Elliott N, Melville L, Beddoe T, Helbig KJ. Crocodilepox Virus Evolutionary Genomics Supports Observed Poxvirus Infection Dynamics on Saltwater Crocodile (Crocodylus porosus). *Viruses. 2019 Dec 2;11(12):1116. doi: 10.3390/v11121116. PMID: 31810339; PMCID: PMC6950651.*

Sekara, V., Stopczynski, A., & Lehmann, S. (2016). Fundamental structures of dynamic social networks. *Proceedings of the national academy of sciences, 113(36), 9977-9982.*

Stewart, J. D., Barroso, A., Butler, R. H., & Munns, R. J. (2018). Caught at the surface. *Ecology, 99(8), 1894-1896.*

van Dam, E. R., & Kooij, R. E. (2007). The minimal spectral radius of graphs with a given diameter. *Linear Algebra and its Applications, 423(2-3), 408-419.*

Wang, Y., Cao, J., Alofi, A., Abdullah, A. M., & Elaiw, A. (2015). Revisiting node-based SIR models in complex networks with degree correlations. *Physica A: Statistical Mechanics and its Applications, 437, 75-88.*

Part 4:
Advanced Applications

Part 4 introduces more advanced algorithms to wrangle network problems, including graph neural networks, vertex clustering, Bayesian networks, ontology web language, subgraphs mining, and graph databases. The problems tackled in this part include the clustering of social network vertices by demographic and network structure factors, understanding the evolution of substance misuse, mining causal pathways related to student learning outcomes, creating gene ontologies, comparing the language classifications of Nilo-Saharan languages, mapping food webs, and creating movie databases with a variety of relationships.

Of note in *Part 4* are *Chapters 13* and *14*. *Chapter 13* combines concepts from throughout this book to construct and analyze data related to Ebola outbreaks in the Democratic Republic of Congo. *Chapter 14* presents cutting-edge network science applications, including quantum network science, network analysis of deep learning architectures, higher-order structuring of networks, and hypergraphs with a focus on medical, environmental, image, and language data applications.

This part has the following chapters:

- *Chapter 9, Machine Learning for Networks*
- *Chapter 10, Pathway Mining*
- *Chapter 11, Mapping Language Families – an Ontological Approach*
- *Chapter 12, Graph Databases*
- *Chapter 13, Putting It All Together*
- *Chapter 14, New Frontiers*

9

Machine Learning for Networks

In this chapter, we'll consider **machine learning** (**ML**) models typically used on relational data and their applications within network science. While many network-specific tools provide good insights into network structure and prediction of spread across a network, ML tools allow us to leverage additional information about individuals in the network to construct a more complete view of relationships, spreading processes, and key outcomes related to the network or its individuals. We'll consider friendship networks and metadata associated with individuals and their connections to other individuals to explore ML on networks.

We'll first return to network construction based on shared activities and traits of individuals, move on to clustering based on both network and metadata features, and finally predict individual and friendship network outcomes based on networks and their metadata. You'll learn how to combine network metrics with metadata and how to build several types of ML models using network data, upon which we will build in the remaining chapters of this book. Let's dive into some friendship networks and their metadata.

Specifically, we will cover the following topics in this chapter:

- Introduction to friendship networks and friendship relational datasets
- ML on networks
- SDL on networks

Technical requirements

The code for the practical examples presented in this chapter can be found here: https://github.com/PacktPublishing/Modern-Graph-Theory-Algorithms-with-Python

Introduction to friendship networks and friendship relational datasets

In this section, we'll consider a friendship network based on student behavior factors to form a network. We'll then apply **unsupervised learning** (**UL**) methods, namely clustering, to group individuals into friendship groups to compare performance before and after adding extra network structural information.

Friendship network introduction

Let's consider a group of classmates in a small school with enrollment based on age and geography, as is common in the United States. Classmates may participate in the same extracurricular activities, such as sports teams, the school paper, or a concert band. They may also study together, share meals, or get together to hang out on weekends. Some may form a core group of friends who take some of the same classes, participate in the same extracurriculars, study together, and hang out together outside of school-related activities. Strong social ties such as these often form an integral source of social support and lasting social relationships. These tend to be very important to individual life decisions and outcomes, particularly in adolescence and early adulthood, where peers play an important role in psychosocial development.

Other groups of friends may only study together or play on the same team, with few other shared interests or interactions. Weak social ties such as these also play an important role in society, connecting individuals with a wide range of resources across a community and exposing young people to a wider variety of viewpoints and new ideas. Social change often comes from weak ties across diverse communities, such as playing on the same sports teams, sharing classes in school, and participating in religious activities. While weak social ties often don't provide strong social support, they serve a bridging function within networks and can introduce individuals to others who will become strong social ties.

In our first friendship network, we'll consider both weak and strong social ties. Strong social ties mainly occur within a group of seven friends who mostly play on the same team, share some classes, study together, and play sports before school and at weekends:

Figure 9.1 – An illustration of a group of boys playing basketball before school

Figure 9.1 shows three of the strong-social-tie boys playing basketball before school. We'd expect ideas, behaviors, and communicable diseases to spread quickly through this part of the network, as this group spends most of its time together.

In contrast, weak social ties within the network consist of occasional interactions that might include core courses or one shared interest that brings individuals together for short periods of time, such that they recognize each other and might know something about fellow students but probably don't know much about other students' interests, home life, or aspirations:

Figure 9.2 – An illustration of students in the same classroom for a
course who may not interact outside of the classroom

Figure 9.2 shows students in the same classroom who may not interact outside of that single class. Weak social ties such as these expose students to different ideas, different interests, seasonal flu, and more but have less influence on an individual than on the group of individuals as a whole.

In this chapter, we'll infer groups of students likely to share strong social ties through UL algorithms on both network metrics and metadata related to student demographics. We'll also analyze social network risk on randomly generated networks to understand different epidemic risks for different types of networks through **supervised learning** (**SL**) with GNNs. Let's explore our initial dataset a bit before diving into some analytics.

Friendship demographic and school factor dataset

In this chapter, we'll mainly work with a dataset containing information about a group of 25 students who are connected by many different lifestyle factors: team membership, casual workouts, weekend sports activities, game attendance, and homework study group membership. Demographic and socioeconomic factors, as well as class assignments, also connect these students by registration in four elective courses, gender, neighborhood of residence, and prior attendance at one of two local junior high schools.

This dataset was derived from Farrelly's secondary school diary over the course of a month in her freshman year. Farrelly herself is individual #7. To create a weighted network, we'll sum up connections across factors between pairs of students. This will give us an approximation of which students are most connected to each other. We'll first explore clustering to discern friendship groups.

Let's see how we can cluster this network based on metadata alone before we move into clustering on both metadata and network metrics.

ML on networks

Now that we have explored friendship data a bit, let's see how clustering algorithm performance varies depending on whether or not we include structural information about the network. We'll start by considering just student factors.

Clustering based on student factors

For our first attempt at clustering, we'll focus on the dataset itself, which contains metadata regarding student demographics and social activities. One of the simplest clustering algorithms is *k-means clustering*, which partitions data iteratively to minimize within-cluster variance and maximize between-cluster variance. This means that students clustered together have more in common with students in that same cluster than with students in other clusters. K-means clustering is a simple algorithm that works well in most cases. However, one needs to specify the number of expected clusters, which is typically not known ahead of time. We'll use a cluster size of 3 and assess model fit; in addition, we'll restart the algorithm five times to ensure that we have an optimal three-cluster solution regardless of algorithm start point and random error.

> **Important note**
>
> If you are on a Windows machine, you may get a warning that does not impact results; some of the packages on `scikit-learn` are not updated with the new Windows operating systems in mind. New releases of operating systems and updates to package dependencies tend to trigger these warnings.

Let's dive into the k-means clustering code with `Script 9.1`:

```
#import packages needed
import pandas as pd
from sklearn.cluster import KMeans
import igraph as ig
from igraph import Graph
import numpy as np
import os
```

```
#import file
File ="<YourPath>/Friendship_Factors.csv"
pwd = os.getcwd()
os.chdir(os.path.dirname(File))
mydata =
    pd.read_csv(os.path.basename(File),encoding='latin1')

#k-means model
X=mydata[mydata.columns.drop('Individual ID')]
km=KMeans(n_clusters=3,init='random',n_init=5)
km_model=km.fit_predict(X)

#explore k-means model
km_model

#add to dataset as first solution
km_1=np.array(km_model)+1
mydata['km_1']=km_1
```

The clustering results suggest that the three-cluster solution is a good fit. One cluster group (*#0*) includes individuals 1-7 and individual 10; this group mostly does homework together, attends games, works out together on weekends, and plays on the same team. Cluster *#0* is characterized by a tight-knit group of friends who share many of the same activities and are near each other most of the week. We'd be concerned about an epidemic starting and spreading with this group. Likely, they share the same protective behaviors, such as healthy eating, regular physical activity, and social engagement. However, an infectious disease or risk behavior that might lead to physical injury (trying a dangerous take on a sports move, taking dares…) is a concern, as the behavior is likely to spread through the entire group of friends.

Another group (*#1*) includes individuals 8-9, 12-14, 16, 19, and 23-25; these individuals usually share *class 2*, don't work out or play sports together outside of school, don't do homework together, and don't share many other classes. Cluster *#1* is characterized by a lack of involvement and engagement with others in our sample. This group is low risk for both protective behavior and risk behavior spreading as they don't have strong social ties to others in our sample. Likely, they wouldn't be influenced or influence others with behavior.

The last group (*#2*) includes individuals 11, 15, 17-18, and 20-22; this group is heterogeneous and includes teammates who don't have much else in common, individuals who share a few classes, and isolated individuals with few connections to others. In general, this group is low risk for epidemic or behavior spread like cluster *#1*; however, they are more active within the sample and may be influenced somewhat by teammates or those with whom they share multiple classes.

Clustering based on student factors and network metrics

Now, let's create a network based on thresholded Pearson correlations, which represents the similarity of activities/classes across individuals by adding to `Script 9.1`:

```
#create network via Pearson correlation
cor=np.corrcoef(X)
cor[cor>=0.5]=1
cor[cor<0.5]=0
X2=np.asmatrix(cor)

#create graph with self-loops removed
friends=Graph.Adjacency(X2)
edge_list=friends.get_edgelist()
self_loop=[]
for i in range(0,25):
    self=(i,i)
    self_loop.append(self)
to_remove=[]
for i in edge_list:
    for j in self_loop:
        if i==j:
            to_remove.append(i)
friends.delete_edges(to_remove)
ig.plot(friends)
```

Running this addition to `Script 9.1` yields a plot of the friendship network, as shown in *Figure 9.3*:

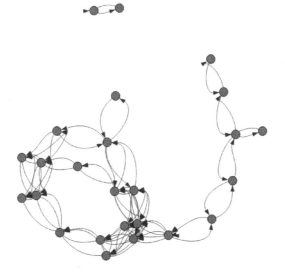

Figure 9.3 – A network plot of the thresholded friendship dataset

Figure 9.3 shows two separate groups, with one very small group consisting of two individuals and a much larger group with sparse and dense connectivity among individuals in the group. We'd expect the degree and PageRank centralities to vary quite a bit among individuals, given the connectivity patterns of our friendship dataset. Let's add to Script 9.1 and append our feature matrix to rerun our k-means analysis, including both demographic factors and two scaled centrality metrics, to see how our clustering changes:

> **Note**
>
> You may find a warning about copying objects; this does not impact the analysis or object.

```
#create scaled metrics and attach to X
d=np.array(Graph.degree(friends))/10
p=np.array(Graph.pagerank(friends))*20
X['degree']=d
X['pagerank']=p

#create new k-means model with graph metrics added
km2=KMeans(n_clusters=3,init='random',n_init=5)
km_model2=km2.fit_predict(X)

#explore new k-means model
km_model2

#add to dataset as first solution
km_2=np.array(km_model2)+1
mydata['km_2']=km_2
```

We can see some changes in our clustering results compared to our initial k-means model. In cluster *#0*, individual 19 is added (a teammate who does homework with the initial *#0* cluster and attends the game). Our initial cluster *#1* shows individuals 8-9, 12, 16, and 23-25; individuals 13, 14, and 19 are no longer assigned to this cluster but other individuals remain. In the remaining cluster, individuals 13 and 14 join our initial cluster, both of whom seem to have more connectivity than initial cluster *#1*, fitting better with cluster *#2* based on centrality metrics. It seems that adding network connectivity metrics improves k-means clustering results, as individuals who may not share every activity but show similar group connections are reassigned to groups that more closely fit their positions within the social network.

Let's now see how we can use a semi-supervised clustering algorithm that we first encountered in *Chapter 5*—spectral clustering—to obtain a semi-supervised solution to our friendship network clustering.

Spectral clustering on the friendship network

As we saw in *Chapter 5*, spectral clustering offers a clustering option to partition either an adjacency matrix or a distance matrix; this can be done as a UL or **semi-SL (SSL)** algorithm. Here, we'll use our correlation matrix from `Script 9.1` to run an unsupervised spectral clustering with three clusters and five initializations (similar to our k-means runs) on our friendship dataset to compare with our k-means results by adding to `Script 9.1`:

> **Note**
>
> Again, you may encounter a Windows warning from scikit-learn or a warning about the graph not being fully connected (assessed via Laplacian, which results in a different approach to the clustering than would be run for a fully connected network). Neither of these warnings will impact the result.

```
#import packages needed
from sklearn.cluster import SpectralClustering
from sklearn import metrics

#perform spectral clustering and attach to dataset
sc = SpectralClustering(3, affinity='precomputed',n_init=5)
sp_clust=sc.fit(cor)
mydata['sp']=sp_clust.labels_
sp_clust.labels_
```

These results differ significantly compared to the k-means solutions we obtained in the previous subsection. Given that both k-means models consider specific activities and course schedules rather than just a correlation summary, this difference makes sense. The spectral clustering solution focuses solely on network connectivity rather than the factors included in the friendship dataset or a combination of connectivity and factors. In this case, the k-means solutions make more sense given our data— particularly the second k-means solution, which includes network metrics and the original factors.

The selection of unsupervised versus semi-supervised clustering algorithms is highly specific to the task at hand. For very large networks, k-means algorithms have solutions that scale well, and adding network connectivity metrics that scale well may improve k-means solutions without sacrificing efficiency. For problems that involve a pure network connectivity solution, spectral clustering may be preferable, particularly if the factors used to construct the network were not collected or are unknown for a third-party network. However, spectral clustering can also take partially labeled data as input, allowing for SSL that can guide the learning process given what is known already about the data.

Now that we've seen how UL and SSL algorithms can be used on network datasets, let's turn our attention to SL algorithms, focusing on an exciting new type of **deep learning (DL)** algorithm specifically designed to take network datasets as their input.

DL on networks

In this section, we'll consider a new type of DL model called GNNs, which process and operate on networks by embedding vertex, edge, or global properties of the network to learn outcomes related to individual networks, vertex properties within a network, or edge properties within a network. Essentially, the DL architecture evolves the topology of these embeddings to find key topological features in the input data that are predictive of the outcome. This can be done in a fully supervised or semi-supervised fashion. In this example, we'll focus on SSL, where only some of the labels are known; however, by providing all labels as input, this can be changed to an SL setting.

Before we dive into the technical details of GNNs, let's explore their use cases in more depth. Classifying networks themselves often yields important insight into problems such as image features or type, molecular compound toxicity or potential use as a pharmaceutical agent, or potential for epidemic spread within a country of interest given travel routes and population hubs.

Typically, data such as molecules or images is transformed into network structure prior to the network embedding step of GNNs. Within the context of molecular compounds, atoms that share a covalent bound, for instance, are represented as vertices connected by an edge. Each compound, then, results in a unique network based on the molecular structure of that compound. For proteins, amino acids can serve as vertices, with connections existing between amino acids sharing a bond (such as a cysteine bridge resulting from a disulfide bond).

When screening potential compounds for use in pharmaceutical development, we often want to predict if the compound might have toxic effects. Using known databases of toxic compounds and compounds with no toxic effects, we can develop a GNN to predict the toxic effects of new compounds in development based on the molecular structures of the new compounds, given what we know about molecules that are known to be or not to be toxic. This allows for quick screening of potential new drugs for toxicity prior to animal or human trials.

GNNs are also able to learn vertex labels given an input network, which is the focus of this chapter. For instance, within a crime or terrorism network, we may wish to identify potential leadership within the network given some knowledge of leaders and non-leaders from collected intelligence data. Incomplete information is common within intelligence data, and learning from what is known can be valuable in identifying key players in the network who are not known and who may be difficult to identify from informants or undercover agents. Since vertex prediction involves a network that has been constructed, we typically skip to the embedding steps of the GNN rather than wrangle the data. However, it might be necessary to add vertex labels to the graph to denote known information about leadership structure in the network.

Edge learning with GNNs mirrors vertex learning, typically through the use of an existing network with complete or incomplete information about edge properties (such as communication frequency or importance across members of a terrorist network that might involve coordinating a terrorist attack or recruiting new members in a geographic region). We embed the edges rather than the vertices in this case before proceeding with the GNN training.

Now that we know a bit about problems we can tackle with GNNs, let's learn more about the architecture and mathematical operations used to build a GNN.

GNN introduction

GNN construction involves a few key steps. In the prior subsection, we mentioned data transformation as a potential first step. GNNs require a network or tensor of networks as input to the embedding step of the algorithm, so data must contain network-structured data and some outcome label associated with the networks themselves or edges/vertices in the network of interest. Some data engineering may be required to wrangle image(s), molecule(s), or other data sources into network structures. In the prior subsection, we overviewed how molecule or protein data can be transformed into a network structure. Many common types of data have standard transformation methods to transform them into network data; for example, in prior chapters, we've transformed spatial and time series data into network structures that could be used as input for a GNN.

Once our data exists in a network structure with a set of labels for networks, edges, or vertices, we're ready to embed the relevant structures at a network, edge, or vertex level. Embeddings aim to find a low-dimensional representation of relevant network geometry at the level of embedding (network, edge, or vertex). They can also include other relevant information, such as other attributes of networks, edges, or vertices. Sometimes, it's advantageous to create these embeddings manually to include both relevant network structure and attribute information. For instance, in our friendship network, we have data on many activities in which individuals participate; we may wish to create an embedding that captures not only the network centrality metrics but also the activity participation for individuals represented as a network vertex. In our k-means example, including both types of information (network structure and collected activity data) improved k-means performance in finding groups we hypothesized to exist.

Many GNN packages in Python, such as PyTorch (which we will use later in the section), have functions that summarize network properties at the network, edge, and vertex level to create an automatic embedding at a specified dimensionality. How we embed data prior to GNN training greatly impacts results, so this step is important to consider when building a GNN. Even with package functions such as the PyTorch one that we'll use, specifying a dimensionality impacts algorithm performance. We don't want too low of a dimensionality (missing key features relevant to the outcome of interest), but we also don't want too high of a dimensionality (which might include a lot of noise). In practice, this parameter is often optimized through grid search.

Once we have the embedding, we can define the outcomes as target labels. We may need to employ one-hot encoding to transform text labels into a sequence of binary outcomes. Just as other DL algorithms can handle multiclass classification problems, continuous outcomes, or other types of distributions, GNNs can fit many different outcomes of interest. This flexibility makes them ideal for modeling outcomes across network classification/regression problems.

The DL architecture itself is not unique. Readers who are familiar with **convolutional neural networks (CNNs)** will recognize many of the components and backfitting algorithms we'll discuss, as they are identical within the context of GNNs. We start with an input layer with a dimension equal to the embedding dimension, and we end with an output layer with a dimension equal to the number of classes of our outcome (for classification problems, which we'll consider in this chapter). When only the input and output layers exist, **neural networks (NNs)** approximate linear regression, with a learned map between input matrices and output vectors. However, between these layers, we typically include hidden layers that further process features between the input and output layers, as shown in *Figure 9.4*:

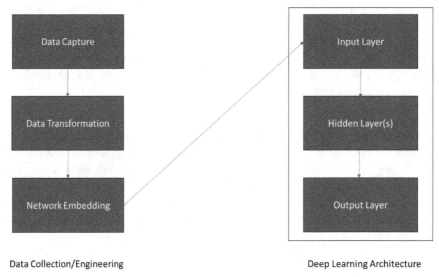

Figure 9.4 – A summary of the GNN life cycle, including data engineering and DL architecture steps

Hidden layers refine the topological maps between the input and output layers, often pooling topological features found during the training process to feed into the next hidden layer. For small networks and small samples of networks, the number of hidden layers should be small to maintain the stability of the solution and obtain good performance. For larger networks or sets of networks, more hidden layers can be added to improve performance without encountering instability of solutions and fit.

Hidden layers typically employ a non-linear mapping function, called an *activation function*, between the input layer and output layer connected to that specific hidden layer. In practice, only a few activation functions are common, including the *ReLU function*, which returns 0 for negative or zero input values and the input value for positive input values.

Convolution layers, also commonly used in hidden layers, apply a filter function (typically a kernel) to the input layer to transform it through the defined kernel function. Typically, a convolution layer will reduce the dimensionality of the matrix or tensor, so zero padding to maintain dimensionality may be used to avoid the problem of transforming layers in a way that limits the number of layers possible given the dimensionality of the output. For small datasets, such as the one we consider, this is not necessarily a problem, as shallow networks tend to be the only stable GNNs that we can train given the limited data size.

The theory of building effective architectures is beyond the scope of this book, and readers interested in DL who do not have a background can obtain this knowledge by reading through the references provided at the end of this chapter.

Once an architecture is defined (either through expert guessing or, again, grid search to optimize architecture), we must fit the parameters connecting nodes in each layer and across layers (called *weights*). There are many options to do this, and it's possible to define custom fitting algorithms. However, we'll focus on the two most common options within the PyTorch package used in our example: **Adam optimizers** and **stochastic gradient descent (SGD)**.

SGD fits weights between nodes within and across layers by exploring the gradient function defined on the NN in much the way that gradient boosting fits a linear regression model. A *learning rate* is defined to control the exploration of the gradient function. A steeper learning rate fits a model more quickly but may not find global minima or maxima. One caveat of SGD is that the algorithm can get stuck in local optima, resulting in lower accuracies of results than what is possible given the input data, mapping functions between layers, and the outcome data. It also tends to be slower, requiring more algorithm iterations and potentially more processing power to even fit the model.

Adam optimizers allow for flexible learning rates for different weights between nodes, leading to faster model fits and avoiding local optima by allowing the learning rate to adjust to the local gradient landscape. Adam also allows for decay rates, further customizing local learning of weights. Many Adam optimizers have evolved since the initial Adam optimizer was developed, and it's likely more will be developed for GNNs and other DL architectures. One drawback is that Adam optimizers tend to be memory-intensive. When training large GNNs, it may be preferable to use SGD to avoid memory issues during training.

In practice, it's difficult to know which optimizer is best for fitting the weights of the defined architecture, and grid search, again, is typically employed for industry GNN models to optimize this choice. Once a fitting algorithm is selected, a predefined (again, usually optimized by grid search) number of iterations is run, or the algorithm runs until meeting a stopping criterion. Adam optimizers tend to converge more quickly than SGD optimizers, but performance can vary depending on the data and architecture.

Now that we understand a bit about the building blocks of GNNs, let's explore an example using an open source sports network consisting of students assigned to two different teachers (our outcome of interest).

Example GNN classifying the Karate Network dataset

For our example, we'll predict vertex-level attributes in a common open source network: Zachary's `Karate Network` dataset. This dataset consists of 34 individuals connected by 78 edges in a karate training network who ended up splitting between an administrator and one of the instructors when a conflict between the administrator and instructor occurred. One of the primary tasks for vertex classification and learning problems on this network is to predict which individuals ended up siding with which person in the conflict (the administrator or the instructor). We will predict vertex labels through a semi-supervised GNN model approach.

We'll first install the needed packages and import our dataset. If you don't have the necessary packages installed, please install them on your machine prior to running our code. We've provided this step as an option in `Script 9.2`:

```
#install packages if you have not installed them on your machine
#!pip install dgl
#!pip install torch

#import packages
import dgl
import dgl.data
import torch
import torch.nn as nn
import torch.nn.functional as F
import itertools
from dgl.nn import SAGEConv

#import Karate Club dataset with instructor/administrator labels
dataset = dgl.data.KarateClubDataset()
num_classes = dataset.num_classes
g = dataset[0]
```

To embed our vertex data, we'll use PyTorch's default embedding algorithm with a dimensionality of 6. Anything that is in the 4-6 dimension range should work reasonably well, given the size of our network. Let's add to `Script 9.2` to embed our vertices:

```
#embed vertices with a dimension of 6
vert_em = nn.Embedding(g.number_of_nodes(),6)
inputs = vert_em.weight
nn.init.xavier_uniform_(inputs)
```

This piece of the script should output embedding vectors for each vertex in our network. Now that we have our vertices embedded, we can create our labels. Given that we wish to demonstrate a semi-supervised approach, we'll feed our network information on six vertices (1, 3, 5, 12, 15, and 32). You can play around with this part of the script to see how fewer or more vertices impact the performance and stability of our chosen architecture. Let's add the label information by adding to Script 9.2:

```
#obtain labels and denote available labels for GNN learning
#(here: 1, 3, 5, 12, 15, 32)
labels = g.ndata['label']
labeled_nodes = [1, 3, 5, 12, 15, 32]
```

Next, we'll need to build our GNN architecture and define training parameters. Many of the papers and tutorials on GNNs using this dataset employ a very shallow network architecture and Adam optimizers. For the sake of comparison and demonstration of other options for building GNNs, we'll use two hidden layers instead of one (including convolution layers coupled with ReLU functions), employ small layers (eight and six nodes, respectively for hidden layers), an SGD fitting algorithm (with a learning rate of 0.01 and a momentum driving the algorithm of 0.8, which is close to the default value), and 990 iterations. Many examples that exist online use Adam optimizers and a single hidden layer with more nodes than our architecture, allowing for fewer training iterations. However, for larger network vertex-label prediction problems, a more complex architecture is likely to perform better, so we will show a way to include more hidden layers and a way to use a different fitting algorithm than Adam. Let's define our architecture and fit our weights by adding to Script 9.2:

```
#build a three-layer GraphSAGE model
class GraphSAGE(nn.Module):
    def __init__(self, in_feats, h_feats1, h_feats2,
        num_classes):
        super(GraphSAGE, self).__init__()
        self.conv1 = SAGEConv(in_feats, h_feats1, 'mean')
        self.conv2 = SAGEConv(h_feats1, h_feats2, 'mean')
        self.conv3 = SAGEConv(h_feats2, num_classes,'mean')

    def forward(self, g, in_feat):
        h = self.conv1(g, in_feat)
        h = F.relu(h)
        h = self.conv2(g, h)
        h = F.relu(h)
        h = self.conv3(g, h)
        return h

#6 embedding dimensions as input,
#a hidden layers of 8 and 6 nodes, and 2 classes to output
net = GraphSAGE(6,8,6,2)
```

```
#GNN training parameters
optimizer=torch.optim.SGD(
    itertools.chain(
        net.parameters(), vert_em.parameters()),
        lr=0.01, momentum=0.8)
all_logits = []

#train GNN
for e in range(990):
    logits = net(g, inputs)
    logp = F.log_softmax(logits, 1)
    loss = F.nll_loss(logp[labeled_nodes],labels[labeled_nodes])
    optimizer.zero_grad()
    loss.backward()
    optimizer.step()
    all_logits.append(logits.detach())
    if e % 90 == 0:
        print('In epoch {}, loss: {}'.format(e, loss))
```

You should see the loss function (logistic regression link function here) that decreases across iterations as your output. Typical accuracies from GNN architectures fall into the 95%-100% range. Because this dataset is small and our architecture is large, your output accuracies may vary quite a bit between runs of the algorithm. This has to do with random sampling within the fitting steps of the algorithm and the coarseness of the underlying gradient landscape. Let's add to Script 9.2 to find our accuracy:

```
#obtain accuracy statistics
pred = torch.argmax(logits, axis=1)
print('Accuracy',(pred == labels).sum().item() / len(pred))
```

Our run of the algorithm gives an accuracy of ~97% in this run of the algorithm. That is on par with the performance of other GNN architectures. However, don't be surprised if your accuracy is significantly lower in one or more runs of the script, as we don't have a large enough sample size to fit this type of architecture. Changing the embedding dimension, architecture, and training algorithm parameters will impact accuracy, and interested readers are encouraged to revise Script 9.2 as a way to see how different choices impact accuracy and stability of fit.

In general, GNN classifiers work much better and show better stability on larger networks and with more labels fed into semi-supervised usage. The Zachary Karate Network dataset is small enough that other methods are recommended to classify the network. However, learning labels on a huge social network (such as those found on social media platforms) or a large geographic network (such as a United States city network with connections defined by roads connecting cities larger than 50,000 people) would result in a more stable GNN solution, and it would be possible to create a very deep architecture. However, to fit these large models, we often need a cloud computing platform, as the large datasets and large number of iterations can be difficult on a laptop.

GNNs have shown great promise in network-based classification problems in many different fields, and it is likely that they will continue to evolve and solve pressing problems with large networks and collections of networks. However, cloud computing solutions are often needed, and this requires expertise working with data and Python notebook solutions on the cloud computing platform used to store data and fit the GNN.

Summary

In this chapter, we considered several use cases of ML algorithms on network datasets. This included UL on a friendship network through fitting k-means and spectral clustering. We considered k-means clustering on both the original dataset of activities in which individuals participated and the original dataset, with added network metrics to improve clustering accuracy. We then turned to SL and SSL on networks and collections of networks through a type of DL algorithm called GNNs. We accurately predicted the labels of individuals in Zachary's Karate Network dataset through a shallow GNN and compared results with other existing solutions to this network classification problem. In *Chapter 10*, we'll mine educational data for causal relationships using network tools related to conditional probability.

References

Acharya, D. B., & Zhang, H. (2021). *Weighted Graph Nodes Clustering via Gumbel Softmax.* arXiv preprint arXiv:2102.10775.

Bongini, P., Bianchini, M., & Scarselli, F. (2021). Molecular generative graph neural networks for drug discovery. *Neurocomputing, 450, 242-252.*

Fan, W., Ma, Y., Li, Q., He, Y., Zhao, E., Tang, J., & Yin, D. (2019, May). Graph neural networks for social recommendation. *In The World Wide Web Conference (pp. 417-426).*

Hartigan, J. A., & Wong, M. A. (1979). Algorithm AS 136: A k-means clustering algorithm. *Journal of the Royal Statistical Society. Series C (Applied Statistics), 28(1), 100-108.*

Imambi, S., Prakash, K. B., & Kanagachidambaresan, G. R. (2021). PyTorch. *Programming with TensorFlow: Solution for Edge Computing Applications, 87-104.*

Kumar, V. (2020). *An Investigation Into Graph Neural Networks (Doctoral dissertation, Trinity College Dublin, Ireland).*

Labonne, M. (2023). *Hands-On Graph Neural Networks Using Python: Practical techniques and architectures for building powerful graph and deep learning apps with PyTorch. Packt Publishing Ltd.*

Liang, F., Qian, C., Yu, W., Griffith, D., & Golmie, N. (2022). Survey of graph neural networks and applications. *Wireless Communications and Mobile Computing, 2022.*

Mantzaris, A. V., Chiodini, D., & Ricketson, K. (2021). Utilizing the simple graph convolutional neural network as a model for simulating influence spread in networks. *Computational Social Networks, 8, 1-17.*

Min, S., Gao, Z., Peng, J., Wang, L., Qin, K., & Fang, B. (2021). STGSN—A Spatial–Temporal Graph Neural Network framework for time-evolving social networks. *Knowledge-Based Systems, 214, 106746.*

Ng, A., Jordan, M., & Weiss, Y. (2001). On spectral clustering: Analysis and an algorithm. *Advances in neural information processing systems, 14.*

Scarselli, F., Gori, M., Tsoi, A. C., Hagenbuchner, M., & Monfardini, G. (2008). The graph neural network model. *IEEE transactions on neural networks, 20(1), 61-80.*

Wieder, O., Kohlbacher, S., Kuenemann, M., Garon, A., Ducrot, P., Seidel, T., & Langer, T. (2020). A compact review of molecular property prediction with graph neural networks. *Drug Discovery Today: Technologies, 37, 1-12.*

Wu, Z., Pan, S., Chen, F., Long, G., Zhang, C., & Philip, S. Y. (2020). A comprehensive survey on graph neural networks. *IEEE transactions on neural networks and learning systems, 32(1), 4-24.*

Zachary, W. W. (1977). An information flow model for conflict and fission in small groups. *Journal of Anthropological Research, 33(4), 452-473.*

Zhang, L., Xu, J., Pan, X., Ye, J., Wang, W., Liu, Y., & Wei, Q. (2023). Visual analytics of route recommendation for tourist evacuation based on graph neural network. *Scientific Reports, 13(1), 17240.*

Zhou, J., Cui, G., Hu, S., Zhang, Z., Yang, C., Liu, Z., ... & Sun, M. (2020). *Graph neural networks: A review of methods and applications. AI open, 1, 57-81.*

10

Pathway Mining

In this chapter, we'll explore pathway mining, where we use network science and reasoning algorithms to uncover paths that exist within sequential data. Pathways to outcomes are common in both the medical field, where disease progression often follows a pathway from one disease state to another, and the educational field, where course material often builds on prior course material within a degree program such as law or medicine. We'll consider a simulated example of medical courses leading to student success or failure in a hypothetical medical school to understand which courses may require extra support for struggling students to ensure ultimate program success.

By the end of this chapter, you'll understand how to spot problems involving pathways to an outcome of interest, apply advanced reasoning algorithms to find likely pathways within a dataset and interpret the results to intervene at key points in the pathway to a given outcome of interest. We'll consider pathway mining within the context of education, but many problems in the real world involve pathways. Let's explore some of these scenarios in more depth.

Specifically, we will cover the following topics in this chapter:

- Introduction to Bayesian networks and causal pathways
- Educational pathway example
- Analyzing course sequencing to find optimal student pathways to graduation

Technical requirements

You will require Jupyter Notebook to run the practical examples in this chapter.

The code for this chapter is available here: https://github.com/PacktPublishing/Modern-Graph-Theory-Algorithms-with-Python

Introduction to Bayesian networks and causal pathways

Many outcomes of interest across industries involve a sequence of events or choices to reach the outcome of interest. The arrival of packages depends on the safe arrival of packages at each relay point between shipping and arrival. Machinery failure can involve **single points of failure (SPOFs)** or cascades of failure, in which multiple parts of the machine fail before the machinery itself fails.

Consider the pathway to drug addiction. First, a person must be in a situation where drug use occurs—through friends, through family, or through a new social group. Then, a person must try an addictive substance. Then, they must like the substance enough to continue using the substance frequently enough to reach a point of physical or psychological dependence on the drug. *Figure 10.1* shows this sequence of steps in a diagram format:

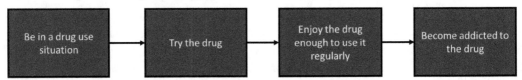

Figure 10.1 – A sequential progression of events leading to drug addiction

Figure 10.1 looks a lot like a directed network, where each situation is a vertex and each step in the pathway is a directed edge. Each edge representing a progression step might be weighted by a probability of progression from one vertex to the next. Let's consider a population of adolescents at high risk of trying a new type of drug and some cross-sectional data on populations at each stage of use that a researcher has collected on the population to determine risks at each step. Let's say that all adolescents have been in a situation where they could try the drug, but only 30% of those who could try it actually do try it. Of those who try the drug, only 20% like it enough to continue using the drug. However, of those who do continue using it, 40% will become dependent on the drug. *Figure 10.2* summarizes this information:

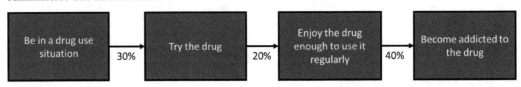

Figure 10.2 – A pathway to drug addiction with probability of transition at each step in the pathway

Figure 10.2 shows a chain of probabilities as drug use progresses through different stages of use. We can find the probability of reaching each stage by multiplying the transition probability of each step before a given stage. For instance, regular use in this population involves trying the drug (30% chance of doing so) and starting regular use (20% chance of doing so); this gives a 6% chance (0.3 multiplied by 0.2) of an adolescent in this population using this new drug regularly. Given that 40% of these adolescents using the new drug regularly end up dependent on it, we'd expect any given adolescent in this population to have a 2.4% chance of progressing to dependency on the new drug given their environment that is conducive to trying the drug.

Mathematics provides us with a formal way to study these pathways or construct them from a set of data. Let's turn to the mathematical tools we need to formalize this intuition of probabilities across a sequence of events.

Bayes' Theorem

The probability of an event that depends on other events is called **conditional probability**. In probability theory, conditional probability is a measure of the probability of an event occurring, given that another event has already occurred. In *Figure 10.2*, the progression to drug dependence relies on regularly using the new drug, which relies upon trying the drug for the first time, which relies upon being in a situation where people use the drug. While conditional probability doesn't need to involve this many conditional steps, it does involve a prior event that influences the probability of an event of interest occurring.

Going back to our example in *Figure 10.2*, we have a universe where an adolescent is exposed to drug use, represented in *Figure 10.3* as a rectangle:

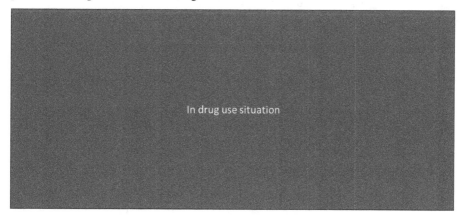

In drug use situation

Figure 10.3 – A universe in which drug use is possible

Figure 10.3 shows an event that is 100% for our group of adolescents; all of them are exposed to this new drug at home, at school, or while they are with friends. However, the event of trying the new drug only occurs 30% of the time, given a new universe that is much smaller (shown in *Figure 10.4*):

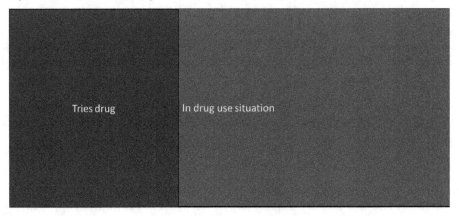

Figure 10.4 – A new universe of trying the drug within our initial universe of being exposed to drug use

Figure 10.4 shows a new universe, where a subset of adolescents try the new drug. Of these adolescents, 20% will regularly use the drug. This creates a new universe, as shown in *Figure 10.5*:

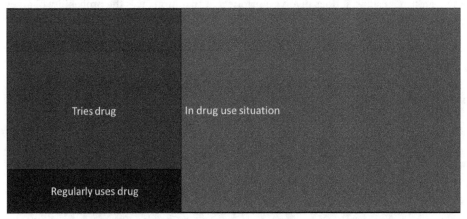

Figure 10.5 – A smaller universe containing adolescents who regularly use the new drug

Figure 10.5 shows a very small universe of adolescents compared to the initial universe of all adolescents at risk of trying this new drug (as we would expect, given that only 6% of adolescents in this population end up using the new drug regularly). Among those who regularly use the new drug, we can split the universe shown in *Figure 10.5* again to obtain the set of adolescents who become dependent on the drug. We won't visualize this universe, as it is too small to visualize well within the small rectangle of regular drug use.

Bayes' Theorem provides a formula for calculating conditional probability by relating prior events' probabilities to the event of interest. We can compute the probability of an event, *A*, given event *B* through the following formula:

$$P(A|B) = (P(A)*P(B|A))/P(B)$$

Here, *P(A|B)* is the probability of event *A* occurring given event *B* occurring, *P(A)* is the probability of event *A* happening, *P(B|A)* is the probability of event *B* occurring given event *A* occurring, and *P(B)* is the probability of event *B* occurring.

To make this more concrete, let's go back to our drug use example. We'll call event *A* trying the new drug and event *B* being present where the drug is used. The probability of event *B* is 100%, as is the probability of event *B* given event *A*. The probability of event *A* is 30%. Let's plug the values into our formula:

$$P(A|B) = (0.3*1)/1$$

This gives us our expected 30%. However, within probability theory, not all events have a probability of 100%, and many calculations will be more complicated as conditional events are added. Within real-world data, we may need to estimate these probabilities with an algorithm given the data we've collected. Let's move on to event chains such as our drug use example.

Causal pathways

Conditional probability can involve more than two events of interest with conditional relationships. **Causal pathways** involve chains of conditional probability that can be relatively short, such as our drug dependence example, or very, very long and complex, such as protein activation pathways contributing to disease risk.

Thanks to Bayes' Theorem, we can chain conditional probabilities together in a piecewise fashion until all conditional probabilities are linked into a final pathway. This works quite well for estimating effect sizes and progression rates for well-known causal pathways. However, many times, we don't know the exact sequence of events leading to an outcome of interest and simply collect a lot of data we think is related to the outcome. To analyze this data, we'll need an algorithm. Fortunately, one exists. Let's dive into Bayesian networks and their application to causal pathway mining.

Bayesian networks

A **Bayesian network** depicts a set of variables and their conditional probabilities as a **directed acyclic graph (DAG)**. Vertices that are not connected by an edge are conditionally independent (not dependent on each other). Vertices connected by an edge are conditionally dependent.

The **chain rule of probability** allows us to construct a Bayesian network as a product of conditional probabilities. Consider events A, B, and C. Returning to our drug use example, event A might be regular use, event B trying the drug, and event C might be around the drug. Our causal chain is thus the following:

$$P(A, B, C) = P(A \mid B, C) * P(B \mid C) * P(C)$$

P(A, B, C) refers to the probability of all events occurring. *P(A|B, C)* refers to the probability of A occurring given that B and C have occurred. *P(B|C)* is the probability of B occurring given that C has occurred. Because these events are usually inferred from data, we need to use an algorithm to estimate the joint probability distribution *P(A,B,C)*. Typically, this is done with an expectation-maximization algorithm through the computation of expected values conditional on the data. Then, the algorithm maximizes the complete likelihood, assuming that the expected values computed are the correct ones. Values are then adjusted again given the likelihood computed in the last step. Once the expected values and likelihood converge, the algorithm stops.

Now that we know the basics of Bayesian networks, let's turn to an educational data example of causal pathways and datasets upon which Bayesian networks can learn.

Educational pathway example

One of the common uses of Bayesian probability and networks is in educational research. Course sequencing involves building upon prior knowledge, and students taking courses conditional on prior knowledge must first obtain that prior knowledge before they can succeed in the course at hand. Prerequisite courses allow professors to require students to take certain courses before taking their courses. Entry to university and then to graduate programs is conditional on successful completion of exams at the previous level of education. Thus, education is a field in which Bayesian networks arise naturally. Let's dive into an example.

Outcomes in education

Many outcomes in education are the culmination of the long sequence of courses students take. For instance, in South Africa and the United States, a student hoping to practice law must take many courses and take a final examination before being able to practice law independently. Passing the examination relies on prior success in coursework and experience gained in internships and other hands-on legal activities. Most medical degrees follow a similar educational approach, with a combination of coursework, practical experience, and final examination culminating in professional status in the field.

Understanding key milestones and turning points within these pathways ensures as many students as possible pass the final examinations to obtain their licenses. However, many courses exist, and not all students take the same courses throughout their education. Medical students may focus on courses most related to their medical subspecialty of interest. Law students may intern within different branches of legal practice to see what type of law they might like to practice. Thus, the data is often incomplete; coupled with small program sizes, this creates a difficult data mining scenario.

Course sequences

One of the major caveats in professional education is course dependencies. For instance, prior to taking a pathology course, medical students typically finish human anatomy. You would expect that success in pathology is at least partially dependent on success in human anatomy. However, it may not be the case that success in pathology depends on success in a human genetics course. It may also be the case that success in these three courses does not influence final examination results (unlikely, but possible).

To make matters even more complicated, certain modules within a course may be more related to student outcomes than other modules. Thus, even a course-level analysis might not be sufficient to pinpoint exactly where students' success or failure usually stems. When looking at these types of real-world problems, it is important to consider the level of analysis under which the pathway is scrutinized.

Antecedents to success

Two approaches are common when collecting data related to student success. One approach, the agnostic one, does not make assumptions about which courses or modules relate most to the outcome. Advantages of this approach include ensuring that all possible data is collected so that any existing relationships may be found. However, the amount of course/module data collected may be large relative to sample size, leading to worse performance of algorithms.

Another approach is to have knowledge about pathways of interest prior to collecting the data. This approach limits the amount of data collected, allowing algorithms to run on a sufficient sample size for good performance; however, if the guess is wrong, the results do not reflect the true pathway that exists in the system.

As we simulate data, we'll take the prior knowledge approach to generate a small dataset to demonstrate how Bayesian networks find pathways in datasets. Let's dive into the dataset simulation and see Bayesian networks in action.

Analyzing course sequencing to find optimal student pathways to graduation

In this example, we'll work with a dataset representing a medical program to understand pathways to the successful completion of a medical degree. In a real medical program, we'd likely include all courses and potentially other factors, such as clinical experiences and research projects required for graduation. However, to run a simple example, we'll assume this data mining has already been done to identify courses related to graduation outcomes.

Introduction to a dataset

Let's imagine a medical program with many courses leading to a final licensing exam. Some courses aren't emphasized by the final licensing exam very much (but are still important to study before entering the field). A handful of courses, though, do show up regularly on the licensing exam, and some build on prior important licensing courses. Let's suppose human anatomy, cellular biology, pathology, microbiology, and neuroscience are five courses that are typically associated with success on the licensing exam. Some material may overlap across the courses—particularly human anatomy, pathology, and microbiology.

We can use Python to simulate student performance in five courses, with three having an overlap of material over the course sequence and two being relatively unrelated to the other three courses, and on a final exam to explore how we would mine for course pathways related to success on a final outcome—our final exam. The numpy package has several useful functions to first generate binomial distributions with different success probabilities (`random.binomial()`) and then to select outcomes from different distributions conditional on other generated probability distributions (using the `where()` clause):

1. Let's see this in action with `Script 10.1`, where we first import our packages and then generate our conditional course distributions for 500 students:

```python
#import needed packages
import pandas as pd
import numpy as np

#create conditional courses that relate to final exam passage
#rates

#course 1, with low passage rates in general
course1=np.random.binomial(1,0.75,500)

#course 2, with low passage rates on the first attempt if course
#1 was failed
course2a=np.random.binomial(1,0.95,500)
course2b=np.random.binomial(1,0.5,500)
course2=np.where(course1>0,course2a,course2b)

#course 3, with passage rates relative dependent on prior
#performance
course3a=np.random.binomial(1,0.95,500)
course3b=np.random.binomial(1,0.65,500)
course3=np.where(course2+course1>1,course3a,course3b)
```

2. We'll then add to `Script 10.1` our final two courses and the dependent performance on the final exam:

```
#create two other courses that are not related to performance on
#final exam
course4=np.random.binomial(1,0.8,200)
course5=np.random.binomial(1,0.85,200)

#create final exam passage rates
passa=np.random.binomial(1,0.95,200)
passb=np.random.binomial(1,0.75,200)
pass_final=np.where(course1+course2+course3>2,
    course3a,course3b)
```

3. Now that we have our course data on pass/fail performance, we can create a data frame containing this data to pass into our pathway mining with a Bayesian network. Let's add this piece to `Script 10.1` to prepare for our pathway mining:

```
Course_Data=pd.DataFrame([course1, course2, course3,
    course4, course5, pass_final],
    index=['Course_1', 'Course_2', 'Course_3',
    'Course_4','Course_5',
    'Pass_Final_Exam']).transpose()
```

Now that we have our dataset, we can turn our attention to the Bayesian network we'll create.

bnlearn analysis

Python has an easy-to-use package to fit Bayesian networks to datasets such as the one we generated in `Script 10.1`: the `bnlearn` package. If you do not have the current version of numpy installed, you'll need to update your numpy version before installing the `bnlearn` package to avoid installation errors. We assume that you have completed this step. Follow the next steps:

> **Note**
>
> There is a `pandas` dependency as well, so we will provide an example of installing a specific version of a package in the following code to deal with the `pandas` versioning dependency. You'll need to restart your Jupyter kernel after this installation.

1. First, we'll install the `bnlearn` package and load it with `Script 10.2`:

```
#install bnlearn package if not already in directory and import
!pip install pandas==1.5.3
!pip install bnlearn
import bnlearn as bn
```

2. Now that we have our package installed, we can fit a Bayesian network to our `Course_Data` dataset using the function's default parameters by adding to `Script 10.2`:

```
#fit Bayesian network
model = bn.structure_learning.fit(Course_Data)
```

The default parameters include a hill-climbing algorithm, which searches the local model space in a greedy fashion (where each step adjusts a single edge), and the **Bayesian inference criterion** (**BIC**) used as a performance measurement (which is a model deviance-based measurement penalized by the number of model parameters).

3. The `bnlearn` package has a nice table summary of dependencies found in the Bayesian network to show which variables are related. Let's add this piece to `Script 10.2` and examine the printed table's results in *Figure 10.6*:

```
#print dependencies
print(model['adjmat'])
```

target	Course_1	Course_2	Course_3	Course_4	Course_5	\
source						
Course_1	False	False	False	False	False	
Course_2	True	False	False	False	False	
Course_3	False	False	False	False	False	
Course_4	False	False	False	False	False	
Course_5	False	False	False	False	False	
Pass_Final_Exam	True	True	True	False	False	

target	Pass_Final_Exam
source	
Course_1	False
Course_2	False
Course_3	False
Course_4	False
Course_5	False
Pass_Final_Exam	False

Figure 10.6 – A summary of our Bayesian network's results

Figure 10.6 shows which relationships exist between different variables. The target is the dependent event, while the source is the prior event upon which the target depends. If the dependency was found to exist, the cell relating the target and source will read as `True`, while if the dependency was not found to exist, the cell will read `False`. Because this is a naïve analysis with respect to the timing of courses, we'll ignore the directionality of the results (such as the final exam not being dependent on any of the courses prior to it according to the directionality of source and target, but courses being found dependent on final exam performance). Typically, Bayesian network analysis is used to find relationships rather than explicit directionality, which can be tested with other statistical methods (discussed later in this chapter).

We do find several of the dependencies we simulated. `Course_2` was found to be dependent on `Course_1`, and the final exam performance is dependent on `Course_1`, `Course_2`, and `Course_3`. While this is not 100% accurate, we did find all four of our dependencies as being within a causal pathway.

4. By adding to `Script 10.2`, we can visualize the DAG we found in our analysis:

```
#plot Bayesian network derived from dataset
bn.plot(model)
```

This plot should show something like *Figure 10.7*, showing the four related variables we simulated as existing within a course success pathway:

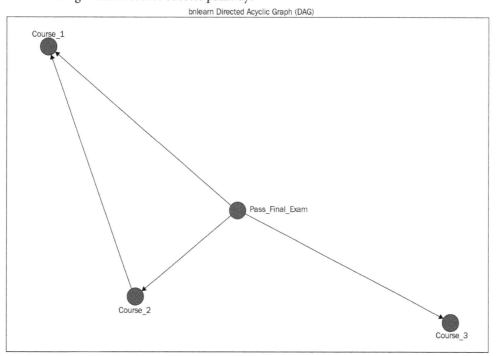

Figure 10.7 – A plot of the Bayesian network's results

Our results are pretty good for a small dataset that we simulated with conditional and random distribution draws. However, Bayesian networks can be very sensitive to algorithms used in fitting. Let's rerun our analysis using an exhaustive search, which scores all possible Bayesian network structures to choose the best model, rather than a hill-climbing algorithm. Note that due to the increased compute time and power needed to fit the model, it is not advised as a fitting algorithm for large datasets or datasets with many variables to explore.

> **Note**
>
> Depending on your system, you may or may not have the following script run to completion on your system. On our machine, the following script took over an hour to run.

`Script 10.3` runs this new fit of a Bayesian network to our three courses designed to depend on prior course performance:

```
#fit Bayesian network
model = bn.structure_learning.fit(Course_Data,methodtype='ex')
```

We can now examine the table of relationships found by our Bayesian network, shown in *Figure 10.8*, by adding to `Script 10.3`:

```
#print dependencies
print(model['adjmat'])
```

target	Course_1	Course_2	Course_3
source			
Course_1	False	True	True
Course_2	False	False	True
Course_3	False	False	False

Figure 10.8 – Bayesian network table of dependencies among three courses we simulated

Figure 10.8 shows that this exhaustive search algorithm does find all three courses to be dependent, as well as the correct directionality of those dependencies—with passing `Course_2` being dependent on passing `Course_1` and with passing `Course_3` dependent on passing `Course_1` and `Course_2`.

Let's now visualize these results with a graph representation of our Bayesian network, adding the following to `Script 10.3`:

```
#plot Bayesian network derived from dataset
bn.plot(model)
```

This yields the output shown in *Figure 10.9*, which shows the dependencies found among the three courses. Note the directionality in *Figure 10.6* matches the dependencies we simulated in our dataset:

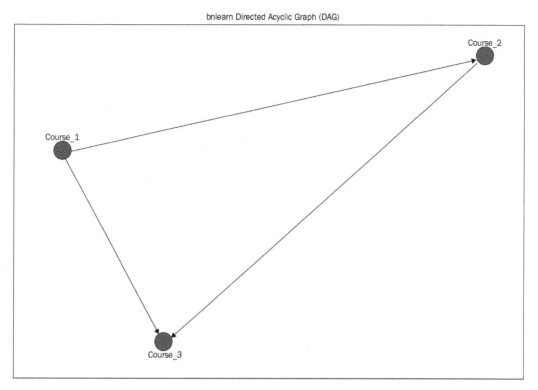

Figure 10.9 – A plot of course pass dependencies among the three
courses simulated to have interdependencies

Note that while the exhaustive search finds the correct causal pathway, it is impractical for most real-world problems, as its run time limits are used on datasets with more variables. Even for our full dataset, the runtime to compute a Bayesian network with exhaustive search is impractical. However, the greedy hill-climbing algorithm was good enough to parse through our data and find likely relationships that exist in the dataset. In the real world, mining for pathways is usually just a first step in understanding a system. Identifying the main components usually suffices for the next steps, which we'll discuss next.

Structural equation models

Once we have an idea of which parts of a pathway may lead to an outcome of interest, we can form a hypothesis regarding the logical steps between these parts. For instance, in our course example, we may know that most students take Course_1 before Course_2 and Course_2 before Course_3. This leads to a hypothesis that Course_1 impacts performance in Course_2 and Course_3 and that Course_2 impacts performance in Course_3. All three courses are assumed to influence performance in the licensing exam. This gives us the hypothesized pathway shown in *Figure 10.10*:

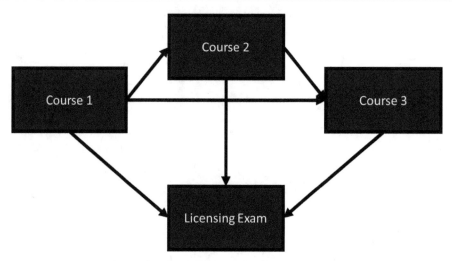

Figure 10.10 – Hypothesized pathway leading to performance in the final licensing exam

Figure 10.10 shows the sequence of events leading to the licensing exam. Performance in prior courses influences performance in future courses and, ultimately, the licensing step. Failure along the way increases the likelihood of future failure, and success along the way increases the likelihood of future success.

Now that we have a hypothesized pathway, we can collect data from another year of students going through this pathway to test our hypothesis. Because we are testing several relationships, we'd want a relatively large sample size. Perhaps 500 students per year take the licensing exam. We may want 2 years' worth of new or historical data on students' pathways to that licensing exam to test our hypothesis with enough statistical power to find effects that exist.

Given that random error influences regression model fitting and statistical results, running six logistic regression models is not ideal; we'd likely find false positives within our set of regression models. However, a handy framework exists to model multiple regression models' fit to causal pathways of interest: **structural equation models (SEMs)**. SEMs provide a framework for fitting causal pathway data within the regression framework, including those with fully measured variables and **latent variables**—those inferred from relationships that exist among measured variables but are not measured directly in the data.

While SEMs are beyond the scope of this book, many frameworks for fitting models exist, and some of the same estimation algorithms and goodness-of-fit statistics that we saw in Bayesian network model fitting are used to fit SEMs. R and Mplus are more commonly used to fit SEMs than Python, but Python includes the `semopy` package to fit SEMs. Note that the types of SEMs are limited compared to the other two software systems and that not all estimation algorithms or goodness-of-fit statistics exist in Python as of 2023. However, if you are interested, I encourage you to explore SEMs as the next step in pathway mining, and references are provided at the end of this chapter if you would like to go further in the field of pathway mining.

Summary

In this chapter, we introduced causal pathways and conditional probability theory through a social science example, building a network-based data mining tool called Bayesian networks. We then simulated data from an educational pathway to implement Bayesian networks in Python. These tools provided a starting point for collecting additional data that could be analyzed to confirm hypotheses constructed from Bayesian networks through a class of models called SEMs. In the next chapter, we'll pivot from causal pathways to look at another niche subfield in analytics: computational linguistics, where we will study languages and their relationships over long periods of time.

References

Gladwin, T. E., Figner, B., Crone, E. A., & Wiers, R. W. (2011). Addiction, adolescence, and the integration of control and motivation. *Developmental cognitive neuroscience, 1(4), 364-376.*

Heckerman, D. (2008). A tutorial on learning with Bayesian networks. *Innovations in Bayesian networks: Theory and applications, 33-82.*

Hoffman, K. I. (1993). The USMLE, the NBME subject examinations, and assessment of individual academic achievement. *Academic Medicine, 68(10), 740-7.*

Hoyle, R. H. (Ed.). (1995). *Structural equation modeling: Concepts, issues, and applications. Sage.*

Igolkina, A. A., & Meshcheryakov, G. (2020). semopy: A Python package for structural equation modeling. *Structural Equation Modeling: A Multidisciplinary Journal, 27(6), 952-963.*

Kaufman, K. A., LaSalle-Ricci, V. H., Glass, C. R., & Arnkoff, D. B. (2007). *Passing the bar exam: Psychological, educational, and demographic predictors of success. J. Legal Educ., 57, 205.*

Mak, K. K., Jeong, J., Lee, H. K., & Lee, K. (2018). Mediating effect of internet addiction on the association between resilience and depression among Korean University students: a structural equation modeling approach. *Psychiatry Investigation, 15(10), 962.*

Meca, A., Sabet, R. F., Farrelly, C. M., Benitez, C. G., Schwartz, S. J., Gonzales-Backen, M., ... & Lizzi, K. M. (2017). Personal and cultural identity development in recently immigrated Hispanic adolescents: Links with psychosocial functioning. *Cultural diversity and ethnic minority psychology, 23(3), 348.*

Ross, S. M. (2014). *Introduction to probability models. Academic Press.*

Scutari, M. (2009). *Learning Bayesian networks with the bnlearn R package. arXiv preprint* arXiv:0908.3817.

Turner, M. E., & Stevens, C. D. (1959). *The regression analysis of causal paths. Biometrics, 15(2), 236-258.*

Violato, C., & Hecker, K. G. (2007). *How to use structural equation modeling in medical education research: A brief guide. Teaching and learning in medicine, 19(4), 362-371.*

Wise, R. A., & Koob, G. F. (2014). The development and maintenance of drug addiction. *Neuropsychopharmacology, 39(2), 254-262.*

Wu, W., Garcia, K., Chandrahas, S., Siddiqui, A., Baronia, R., & Ibrahim, Y. (2021). Predictors of performance on USMLE step 1. *The Southwest Respiratory and Critical Care Chronicles, 9(39), 63-72.*

Mapping Language Families – an Ontological Approach

In this chapter, we'll explore an important area of science called linguistics. **Linguistics** is the study of languages, including word usage patterns, grammar evolution, and the social constructs of language. We'll explore **ontologies**, which relate concepts such as words to each other through defined relationships, and build **language families**, where languages within a family are related to each other across time and geography.

Network science provides us with tools to summarize and compare ontologies and language families to study hypothesized similarities and differences. We'll quantify differences in hypothesized language families at the end of this chapter, though the methodology can apply to ontologies or any other network-based structures derived from linguistic studies.

In this chapter, we will cover the following topics:

- What is an ontology?
- Language families
- Mapping language families

By the end of this chapter, you'll know some of the basic tools used in linguistic-related analyses and be able to compare different networks to understand the key differences between them. Because these tools scale well, you'll be able to tackle even very large ontologies.

Technical requirements

You will require Jupyter Notebook to run the practical examples in this chapter.

The code for this chapter is available here: `https://github.com/PacktPublishing/Modern-Graph-Theory-Algorithms-with-Python`

What is an ontology?

In this section, we'll learn about ontologies and how they can be represented and stored as networks. While this chapter focuses on linguistic ontologies, many other types of ontologies exist, including gene ontologies, briefly reviewed to motivate this section.

Introduction to ontologies

As mentioned in the chapter introduction, an ontology is a set of ideas that are related by some property. Generally, they are used to organize knowledge within a specific discipline. Let's make this a little bit more concrete with an example. Because examples in genetics are more widely used in ontology study, we'll build up the intuition with a biological example before tackling linguistics.

Say we have five genes that are within a regulatory pathway that controls the production of enzymes or polypeptides found in an animal's venom. Often, environmental conditions and geography play a role in the ratio of active compounds found in venom, even within the same species, through regulating gene transcription within the venom glands. Before we start collecting information on these genes and their influence on venom production, we simply have a collection of genes, as shown in *Figure 11.1*:

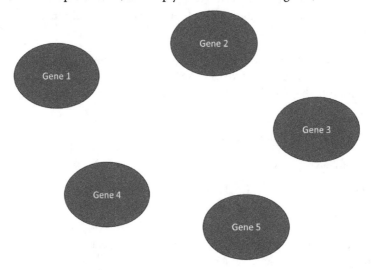

Figure 11.1 – A collection of five genes involved in regulating venom production

To better understand the genes in *Figure 11.1*, we might run an environmental experiment with varying conditions to turn genes on and off and elucidate their relationships with each other—the dependencies that exist within this regulatory pathway. Interdependencies can be represented by connecting each of these gene vertices with an edge. Let's say we ran several experiments and found the following relationships:

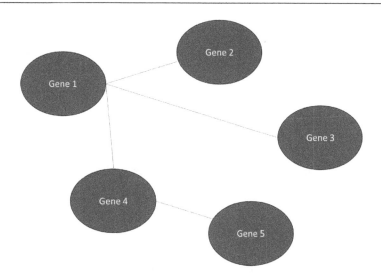

Figure 11.2 – An ontology representing relationships between genes that regulate venom production

Figure 11.2 shows that all four other genes are downstream from *Gene 1*, suggesting that this gene is critically important to the regulation of a critical compound in the animal's venom. Mutations in *Gene 1* likely result in loss of function (and potentially endanger the animal). Visually representing this ontology as a network allows us to easily pinpoint key information within the system we are studying.

However, many organisms' gene regulation pathways are much larger than this example and are hard to visualize in practice—particularly entire gene ontologies, where the organisms' full set of genes is mapped and connected via functional pathways. This means that we'll need to use network analytics tools to study differences in proposed gene regulation pathways or compare gene regulation pathways across organisms that show the same function. For instance, in our venom example, we might be interested in mapping the full gene ontology of a species of sea krait, a member of the elapid family, and comparing its full gene ontology with another elapid species, such as taipan or a black mamba. These species can have upward of 12,000 genes, creating a very large ontology comparison problem.

Let's return to our focus on linguistics and see how language-related ontologies are constructed and can be used to study language differences among dialects, pidgins, or socioeconomic groups—or simply to understand the structure and function of a language.

Consider two words in the English language: *go* and *went*. Both words are verbs. Both have the same semantic meaning with respect to the sentence's subject. However, one is present tense and the other is past tense. All these pieces of information help us understand the functionality of these words within the English language. We can map these relationships to a network structure for each word, as *Figure 11.3* summarizes:

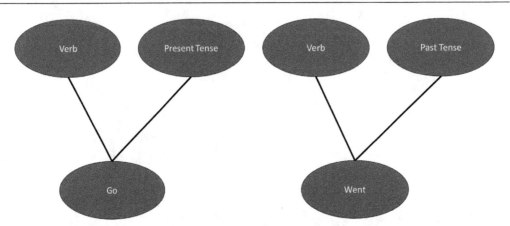

Figure 11.3 – Two networks showing the linguistic properties of the words "go" and "went"

Figure 11.3 summarizes the linguistic properties of our two words: *go* and *went*. Imagine creating networks like this for every word in the English language. We would have very many networks mapping linguistic properties to words, and it would be easy to lose important networks. We'd also use a lot of computing power trying to search through graph collections for words or properties of interest. Let's explore ontology creation further.

Representing information as an ontology

To create an ontology, we need to define concepts/words/terms of interest and the properties that they might share. In our *Figure 11.3* example, we have a set of words with linguistic properties of interest. Once we have these two sets defined (words and linguistic properties, respectively, in *Figure 11.3*'s example), we can create maps between these two sets (the lines shown in *Figure 11.3*). Our word example was a very small example, though. Many times, these sets can include thousands or tens of thousands of items in each of our sets.

To deal with these larger sets of information and the potential relationships that exist between items in them, we need a systematic way to store both sets and their relationships. We'll also need a lot of people to create this very large ontology, preferably with a way to share the information among all of the people working to create this ontology. That's where databases come into play.

Ontology Web Language (**OWL**) database systems organize knowledge networks like the ones shown in *Figure 11.3*. Each word, term, or concept occupies a row within the database that is mapped to whichever properties of interest within the ontology match that word/term/concept, thereby creating a sort of database of databases to store word/term/concept data along with their relationships to key properties of interest in the ontology. From this database, we can search for information, visualize relationships of interest, and mine the entire ontology for overarching themes.

Any type of knowledge can be organized in an OWL database. For instance, we could use an OWL database to organize knowledge about animal taxonomies or evolutionary genetics to quickly search for relationships among colubrid snake species according to genetic analysis results. We could also organize medical terms to create an easy-to-use lookup system for doctors to use when confronted with symptoms or diseases beyond their areas of expertise to facilitate the diagnosis of a patient. We could even organize knowledge about character relationships in long fictional works to provide study guides for a book such as *War and Peace*.

Now that we know the basics about ontologies and the systems in which we can store them, let's turn our attention back to linguistics, where we may wish to study the evolution of languages across language families.

Language families

Language families are groups of languages with a common ancestor that evolved (usually) within a related geographic region. For example, the Romance language family consists of languages that evolved from Vulgar Latin (late-period Latin) in the Southern European region. The Nilo-Saharan language family evolved in the Nile Delta during the time of the Nubian kingdom. There are roughly 150 language families rolled into major language families that exist today, plus several **language isolates** that do not have relationships with other languages (Australian Tiwi, for instance). Language isolates have no genetic ancestors linked to other languages. Campbell gives seven major language families (one per continent), though other researchers classify major language families into other groups of languages.

Language drift and relationships

Language families emerged as different populations of genetic ancestor languages became geographically isolated enough for the ancestor language to evolve regionally. As languages evolve, words may change. Pronunciations may change. Grammar may evolve. In many English **pidgins** (mixes of two or more languages, including English in this case), English has mixed with local languages to create a hybrid language that retains some of the grammar and phrasing of English alongside a local language. In Nigerian English pidgins, English mixes with local languages, such as Igbo, Hausa, or Yoruba. For instance, "*How are you?*" becomes "*How na?*" and "*Please come join me for a meal*" becomes "*Abeg come chop food.*"

Over time, pidgin languages can evolve to the point of being a separate language altogether (a child of English, much the way Middle English evolved from Old English as other languages influenced Old English) or remain a pidgin of the parent language (English, in this case). When the language evolves, it becomes a child branch of the parent language. In the case of English, its widespread use in business and school systems suggests that many child languages will develop over time into separate languages with distinct diction and grammar. Given pidgins in Nigeria and their widespread usage, a Nigerian English child language is probably the most likely child language to develop in the next century or so.

A few driving factors tend to favor language evolution into child languages or pidgins. Commerce between different countries or regions that do not overlap linguistically necessitates means of communication. Pidgins often develop in marketplaces where many speakers of different languages must communicate with each other to buy and sell goods. Common words and phrases might be used over and over during a transaction (such as currency values or amounts). Tricky grammar or articles might be dropped to facilitate understanding between speakers of two different languages.

Empire expansion is another driver of language evolution. The Roman Empire spread Latin throughout much of Europe, influencing languages, particularly those closest to Rome, over centuries. Spain and Portugal are interesting cases, where both Roman influences in Latin and Moor influences in Arabic influenced local dialects. Over time, the Romance languages, including Spanish and Portuguese, evolved out of Latin as the Roman Empire's power waned and other linguistic and cultural influences separated regions that were united in their use of Latin for commerce, schooling, and religion. In fact, the Roman Empire influenced the development of these languages so directly that the term **Romance languages** refers to the Roman Empire!

Most English pidgins that exist today owe their existence to British colonialism, as do different dialects of English spoken across Great Britain, the United States of America, and Australia. Depending on how long English lasts as business's lingua franca, we may see many child languages evolving from English in the next centuries.

A **phylogeny tree** can be used to summarize the evolutionary relationships between ideas, languages, or species over time visually by showing parent and child relationships. This makes the study of evolutionary structure easier, as the visuals provide clues about hypothesized pathways of evolution and a network structure that is easy to analyze for information and potential statistical differences. General ontology structures do not have a hierarchical structure, where one part of the network is ideologically dependent on a prior part of the network. Let's see this in action with the hypothesized evolutionary pathways of Nilo-Saharan languages.

Nilo-Saharan languages

Nilo-Saharan languages include languages spoken in South Sudan, Ethiopia, Sudan, Kenya, Tanzania, Uganda, Chad, Burkina Faso, Cameroon, Mali, the Democratic Republic of Congo, Benin, Algeria, Nigeria, Niger, Egypt, and Libya. Further small language communities exist in several other countries, as well. Some common languages included in this family are Maasai, Songhay, Dinka, Luo, and Kanuri. About 70 million people speak a Nilo-Saharan language, with the number of speakers increasing for some regions and dialects.

The linguistic study of Nilo-Saharan languages dates to at least the 1800s, and many papers in the last decade have debated language drift and relationships of parent languages with their child languages. In this section, we'll examine a few of the proposed hierarchies and then return to analyze the similarities and differences of these hierarchies with Python code in the next section. If you are interested in the papers themselves, you can consult the references at the end of the chapter, particularly the Dimmendaal paper, which summarizes recent debates on parent-child relationships in the Nilo-Saharan family.

Greenberg proposed the following hierarchy in 1963, classifying several language family branches. *Figure 11.4* shows this hierarchy, including children branches off a child branch of the larger Nilo-Saharan family:

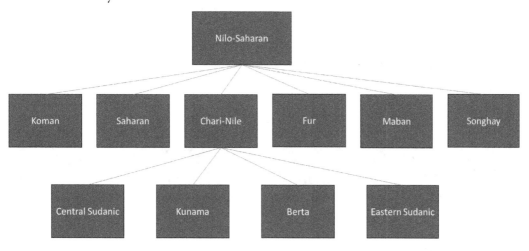

Figure 11.4 – Greenberg's classification of Nilo-Saharan languages

Figure 11.4 shows some subclassifications of Nilo-Saharan languages that are part of the language family. Each of these subclassifications continues to individual languages, totaling roughly 130 languages spoken in this region that originated near the area where the Nile tributaries meet in what was once Nubia.

Since 1963, many papers have further divided these subclasses and added additional subclasses culminating in the Dimmendaal paper that we'll discuss in this section, occasionally creating very complicated and deep network structures. Grammar, diction, phonology (pronunciation of consonants, in this case), and terminology typically dictate where lines are drawn, but languages are often fluid and may overlap somewhat, making classification systems difficult to discern. Let's turn to one of the more recent works on the Nilo-Saharan family of languages.

Dimmendaal's recent work proposes a new hierarchy of the Nilo-Saharan subfamilies, summarized in *Figure 11.5*:

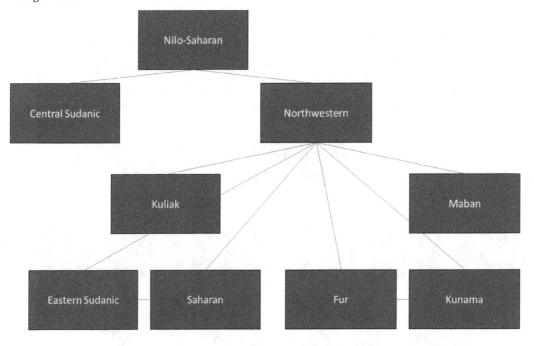

Figure 11.5 – Dimmendaal's hierarchy of Nilo-Saharan subfamilies

Note

In *Figure 11.5*, Dimmendaal's classification splits languages first by region and then by dialect or region. Note also that some language subfamilies appear to have emerged together (such as Fur and Kunama) without the presence of an intermediary family. This is a more complicated structure than Greenberg's classification in *Figure 11.4*.

As language family hierarchies evolve as languages are studied more systematically over time, it is important to note differences that emerge and their potential causes. This shines light on avenues of further linguistic research and can contribute to future language family hierarchy development.

Now that we know a little bit about Nilo-Saharan languages and their hierarchies, let's see how network science can help us understand differences in structure that may be important.

> **Note**
>
> Our example is very limited in size such that it is easy to visualize results and why our analyses may show differences. Working with full OWL databases or full language hierarchies creates much more nuanced and large networks to study. However, the tools for doing so are the same regardless of network size or complexity. Thus, the analyses we're about to begin can be applied to much more complicated linguistic data structures where simply visualizing the differences between OWL databases created by different research groups or different language family hierarchies may not be feasible.

Mapping language families

Let's return to our two examples of Nilo-Saharan language family trees. To examine differences in structure, we'll need to first create these in NetworkX. For the Dimmendaal family tree, we'll break with the tree structure slightly to show the interrelationships that exist for some of the child subfamilies. Let's get started by creating and plotting the Greenberg tree with Script 11.1:

```
#load needed packages
import numpy as np
import networkx as nx
import matplotlib.pyplot as plt

#create the Greenberg Nilo-Saharan language family network
G = nx.Graph()
G.add_nodes_from([1, 11])
G.add_edges_from([(1,2),(1,3),(1,4),(1,5),(1,6),(1,7),(7,8),
    (7,9),(7,10),(7,11)])

#plot the Greenberg Nilo-Saharan language family  network
import matplotlib.pyplot as plt
G.nodes[1]['subfamily'] = 'Nilo_Saharan'
G.nodes[2]['subfamily'] = 'Koman'
G.nodes[3]['subfamily'] = 'Saharan'
G.nodes[4]['subfamily'] = 'Songhay'
G.nodes[5]['subfamily'] = 'Fur'
G.nodes[6]['subfamily'] = 'Maban'
G.nodes[7]['subfamily'] = 'Chari_Nile'
G.nodes[8]['subfamily'] = 'Central_Sudanic'
G.nodes[9]['subfamily'] = 'Kunama'
G.nodes[10]['subfamily'] = 'Berta'
G.nodes[11]['subfamily'] = 'Eastern_Sudanic'
labels = nx.get_node_attributes(G, 'subfamily')
nx.draw(G, labels=labels, font_weight='bold')
```

This script produces a similar representation to *Figure 11.4*. *Figure 11.6* shows the NetworkX plot of the Greenberg Nilo-Saharan language subfamily tree:

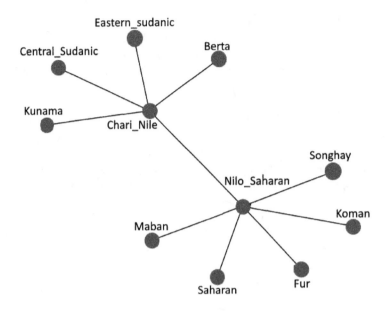

Figure 11.6 – A plot of Greenberg's Nilo-Saharan language subfamily tree

Now that we can see Greenberg's Nilo-Saharan language subfamily tree and its various branches, let's create and plot the Dimmendaal Nilo-Saharan language subfamily tree and its various branches by adding to `Script 11.1`:

```
#create the Dimmendaal Nilo-Saharan language family network
G2 = nx.Graph()
G2.add_nodes_from([1, 9])
G2.add_edges_from([(1,2),(1,3),(2,4),(2,5),(2,6),(5,6),(2,7),
    (2,8),(7,8),(2,9)])


#plot the Dimmendaal Nilo-Saharan language family  network
G2.nodes[1]['subfamily'] = 'Nilo_Saharan'
G2.nodes[2]['subfamily'] = 'Northeastern'
G2.nodes[3]['subfamily'] = 'Central_Sudanic'
G2.nodes[4]['subfamily'] = 'Maban'
G2.nodes[5]['subfamily'] = 'Kunama'
G2.nodes[6]['subfamily'] = 'Fur'
G2.nodes[7]['subfamily'] = 'Saharan'
G2.nodes[8]['subfamily'] = 'Eastern_Sudanic'
G2.nodes[9]['subfamily'] = 'Kuliak'
```

```
labels2 = nx.get_node_attributes(G2, 'subfamily')
nx.draw(G2, labels=labels2, font_weight='bold')
```

This piece of the script produces a plot of Dimmendaal's Nilo-Saharan language subfamily tree, shown in *Figure 11.7*:

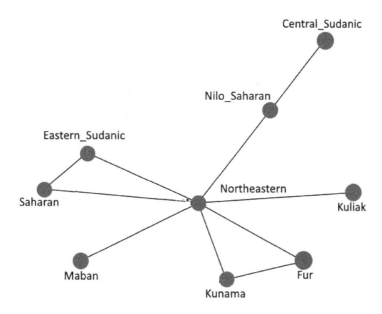

Figure 11.7 – A plot of Dimmendaal's Nilo-Saharan language subfamily tree

Note the connections that exist in Dimmendaal's Nilo-Saharan language subfamily tree between Kunama and Fur, as well as Saharan and Eastern Sudanic. These represent related child subfamilies within the Northeastern branch of Nilo-Saharan languages. From *Figures 11.6* and *11.7*, we can see a similar overall structure, where a lot of shallow, unconnected branches exist. However, Dimmendaal's Nilo-Saharan language subfamily tree adds an extra layer to the tree and includes the related subfamilies without a common ancestor directly begetting them.

Given these differences, we'd expect some centrality metrics to vary between the two language subfamily trees. We'll start with some familiar metrics defined in earlier chapters. Script 11.2 analyzes the differences in average betweenness centrality across our two language subfamily trees:

```
#compare betweenness centrality of language families
gb=nx.betweenness_centrality(G)
print(np.mean(np.array(list(gb.values()))))
gb2=nx.betweenness_centrality(G2)
print(np.mean(np.array(list(gb2.values()))))
```

We obtain average betweenness centralities of 0.131 (Greenberg's) and 0.127 (Dimmendaal's), suggesting that bridging properties are similar across these subfamily trees. Let's assess hubness properties by calculating the average degree centrality by adding to Script 11.2:

```
#compare degree centrality of language families
gd=nx.degree_centrality(G)
print(np.mean(np.array(list(gd.values()))))
gd2=nx.degree_centrality(G2)
print(np.mean(np.array(list(gd2.values()))))
```

We obtain average degree centralities of 0.182 (Greenberg's) and 0.278 (Dimmendaal's), suggesting that hubness properties vary somewhat across these subfamily trees. Given the added layer of depth connecting other layers of the tree in Dimmendaal's subfamily tree structure, we'd expect to see this difference.

Let's turn our attention to a centrality metric especially suited for real-world linguistic data (and other ontology or phylogeny network datasets).

Subgraph centrality measures how each vertex participates within subgraphs on the network by weighting small subnetworks heavily and larger subnetworks less, using the spectra of the network's adjacency matrix, and estimating subnetworks through random walks on the network. This blends the properties of **spectral clustering** (seen in *Chapter 5*) and degree centrality to create a hybrid centrality measurement.

We can see this in action on our two subfamily trees by adding to Script 11.2:

```
#compare subgraph centrality of language families
gs=nx.subgraph_centrality(G)
print(np.mean(np.array(list(gs.values()))))
gs2=nx.subgraph_centrality(G2)
print(np.mean(np.array(list(gs2.values()))))
```

This script shows that average subgraph centrality varies between the two subfamily trees, with Greenberg's average subgraph centrality of 2.478 and Dimmendaal's average subgraph centrality of 3.276. This suggests that the spectra and degree distributions are different across these hypothesized subfamily trees.

While this example only involves two small ontology structures, these types of analyses can be helpful when working with many ontologies or comparing very large ontologies where visualization is not recommended. For example, we may wish to compare the full Nilo-Saharan language trees hypothesized across many different papers. These trees would include hundreds of vertices and possibly dozens of trees. Tracking the differences visually would be quite a challenge!

When we consider linguistic OWL databases, the scale of our analysis grows even further to hundreds of thousands of vertices to compare across OWL databases. At this scale, it is not possible to visually mine the data in a comparison. Network science provides a way to compare the proposed OWL database structures and zero in on differences that exist between those databases to find key differences between the proposed structures. In addition, many of the tools we've overviewed scale well to this size of data analysis, providing an efficient way to find these key differences. However, understanding the differences that exist in etymology theory is beyond the scope of this book, and many OWL databases include storage in proprietary systems or are proprietary themselves.

This chapter has focused on the field of linguistics. However, many types of ontologies and phylogenies exist in other disciplines. Zoology and human genomics often use ontologies to study relationships between genes, environments, and gene expression. These ontologies tend to be large, and the curation of knowledge to create these ontologies can result in significant differences or knowledge gaps between different ontologies attempting to capture the same knowledge. Scalable centrality metrics, such as those we overviewed in a simple-to-visualize example, can pinpoint overall and individual branch differences across these complicated ontologies to pinpoint areas of interest and potential caveats of using one open source ontology versus another.

As knowledge in different fields grows, the methodology to compare ontologies will develop further, and network science will play a significant role in the management, understanding, and usage of ontologies within different disciplines.

Summary

In this chapter, we learned how to represent linguistic data in databases and networks. We compared two Nilo-Saharan language subfamily networks using centrality metrics and introduced subgraph centrality as a centrality metric particularly suited to analyzing ontologies and phylogenies. Finally, we expanded our treatment of ontologies to those outside the field of linguistics and discussed how the tools in this chapter can be used to scale these analyses to very large groups of very large ontologies to compare suitability for use in real-world applications.

In the next chapter, we will learn about graph databases, which can be used to store ontology data such as the data discussed in this chapter.

References

Campbell, L. R. (2018). *How many language families are there in the world?*

Dimmendaal, G. J. (2019). *A typological perspective on the morphology of Nilo-Saharan languages.* In *Oxford Research Encyclopedia of Linguistics.*

Dimmendaal, G. J., Ahland, C., Jakobi, A., and Lojenga, C. K. (2019). *Linguistic features and typologies in languages commonly referred to as 'Nilo-Saharan'. Cambridge Handbook of African Languages*, 326-381.

Done, B., Khatri, P., Done, A., and Draghici, S. (2008). *Predicting novel human gene ontology annotations using semantic analysis. IEEE/ACM transactions on computational biology and bioinformatics, 7*(1), 91-99.

Ehret, C. (2006). *The Nilo-Saharan background of Chadic. Studies in African Linguistics, 35*, 56-66.

Estrada, E., and Rodriguez-Velazquez, J. A. (2005). *Subgraph centrality in complex networks. Physical Review E, 71*(5), 056103.

Gene Ontology Consortium. (2008). *The gene ontology project in 2008. Nucleic acids research, 36*(suppl_1), D440-D444.

Greenberg, J. H. (1957). *Essays in linguistics.*

Greenberg, Joseph H. 1963. *The Languages of Africa*. Bloomington, IN: Indiana University.

Jakobi, A., and Dimmendaal, G. J. (2022). *Number marking in Karko and Nilo-Saharan. Number in World's languages: A Comparative Handbook*, 63-106.

McGuinness, D. L., and Van Harmelen, F. (2004). *OWL web ontology language overview. W3C recommendation, 10*(10), 2004.

Oladnabi, M., Omidi, S., Mehrpouya, M., Azadmehr, A., Kazemi-Lomedasht, F., and Yardehnavi, N. (2021). *Venomics and antivenomics data: Current and future perspective. Archives of Biotechnology and Biomedicine, 5*(1), 026-031.

Souag, L. (2022). *How a West African language becomes North African, and vice versa. Linguistic Typology, 26*(2), 283-312.

Subich, V. G. (2020). *STRUCTURAL AND TYPOLOGICAL ANALYSIS OF THE GENEALOGY OF AFRICAN LANGUAGES.* Арабистика Евразии, (12), 103-117.

Suryamohan, K., Krishnankutty, S. P., Guillory, J., Jevit, M., Schröder, M. S., Wu, M., ... and Seshagiri, S. (2020). *The Indian cobra reference genome and transcriptome enables comprehensive identification of venom toxins. Nature Genetics, 52*(1), 106-117.

Urban, M. (2021). *The geography and development of language isolates. Royal Society open science, 8*(4), 202232.

Wardhaugh, R. (1972). *Introduction to Linguistics.*

Nigerian Pidgin – 20 useful words and phrases (n.d.): https://www.britishcouncil.org/voices-magazine/nigerian-pidgin-words-phrases

12
Graph Databases

In the previous chapters, we've considered networks as data structures that we can analyze to extract insight into data science problems. In this chapter, we'll consider networks as data storage options, linking many pieces of information in a multi-relational way. Many storage options exist but we'll focus on an open source option that integrates well with Python—**Neo4j**.

We will cover the following topics in this chapter:

- Introduction to graph databases
- Querying and modifying data in Neo4j

By the end of this chapter, you'll understand the advantages of graph databases to store network science datasets, be able to visualize graph databases and know how to query them for the quick retrieval of relevant information. You'll feel comfortable modifying tables by insertion and deletion. You'll understand how the tools in our previous chapters can help you query efficiently to find relevant data. Let's get started by exploring the rationale behind graph databases.

Introduction to graph databases

Graph databases (databases that store data in network form) offer many advantages over traditional relational databases. First, graph databases can capture and traverse hierarchical relationships. While relational databases can capture taxonomies, they do so in different columns that are not explicitly linked.

In addition, graph databases capture complex relationships between items or groups of items explicitly by connecting them with edges. This allows for multiway relationships to exist within the database; querying for nearest neighbors, for instance, is much easier when neighbors are connected by an edge and do not require estimation steps to find **Euclidean** or **Manhattan distances** between all items in the database.

Furthermore, graph databases can capture the directionality of relationships between items in the database very easily. In a relational database, the directionality of a single relationship may involve several columns' worth of information to capture that single relationship. Thus, for items with complicated relationships that may be unidirectional, a graph database provides a compact representation of data structure.

Finally, within graph databases, it is possible to mine the dataset for implicit relationships that have not been programmed into the database by using network algorithms to probe for similar patterns in different parts of a network or to infer edges that don't exist based on triadic closure. For many datasets—such as continuously-growing -*omics* datasets, evolving ontologies, or other non-static datasets where knowledge is incomplete—mining the data provides new research avenues.

Now that we know a bit about the advantages of graph databases, let's dive into specific differences between graph databases and relational databases.

What is a graph database?

Graph databases are databases based on network science. Rather than storing data as columns in a spreadsheet (shown in the following figure), as is done in relational databases, graph databases store data as *networks*, with items that would occupy a row represented as a vertex connected to other vertices per relations that exist. Column data is used to either add metadata to the vertex or connect it to other vertices. Graph database queries rely on network science algorithms to traverse the graph, pattern-finding algorithms, and combinations of these two approaches. Thus, queries for graph databases can become quite sophisticated without necessarily leading to a long run-time as subqueries in SQL will require. Efficiency is a big advantage of graph databases. Let's look at an example of supermarket customer feedback related to customer service or products, shown in *Figure 12.1*:

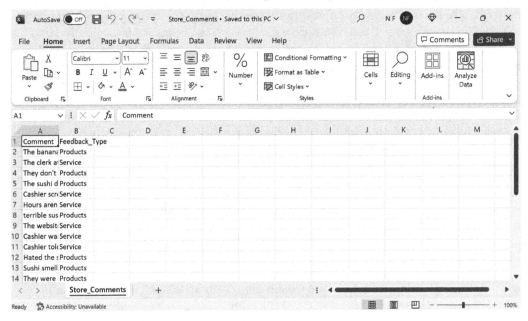

Figure 12.1 – An example of relational database design, where rows contain items with information organized into columns (Used with permission from Microsoft Excel)

Figure 12.1 shows a relational database with many rows of data that exist as columns. However, this data may be better represented as a graph database, where items relate through the columns or use columns as metadata in a queryable way. Perhaps there are dates attached to the comments that could link items by time and use the department related to the complaint as metadata.

While it is difficult to visualize a relational database, visualizing a graph database is much more straightforward. Some graph databases come with built-in visualization tools to zoom into areas of interest or query results. Some allow developers to build on pre-existing visualizations to tailor this feature to their needs. When hovering over edges and vertices, most graph database visualization tools will show metadata associated with a vertex or edge, providing context for the relationships of interest pulled by the query. Cytoscape is one of the most commonly used programs for network visualization at scale; originally, it was constructed to visualize proteomics datasets, where protein interactions are documented across organisms. Here's Cytoscape with demo images of the example graphs: `https://js.cytoscape.org/`.

One of the advantages of Cytoscape is its ability to handle saved igraph networks such that analysis of very large networks can be done in Python and then visualized with Cytoscape, where the network size will not complicate visualization as it will with the igraph plotting functions. In addition, Cytoscape offers a desktop version that should work with Java-based devices and a web interface version for those who do not have access to the desktop version. If you are interested, you are encouraged to download Cytoscape and test out its visualization either with the datasets we use in this book or with your own datasets of interest.

Now that we understand the structure and advantages of graph databases, let's dive into some datasets where a graph database may function better than a relational database with respect to organizing the information and retrieving information of interest.

What can you represent in a graph database?

There are many data science domains that can benefit from graph databases. We'll explore a few of these use cases in depth before diving into coding in one open source graph database. Let's start with an example from epidemiology.

Contact tracing

When a new epidemic starts in an area, epidemiologists often employ a tool called **contact tracing**, where data on each case is collected to connect cases to each other in the hopes of finding the outbreak source, or patient zero. Patient zero often provides insight into the source of the epidemic—whether an animal vector, a laboratory breach, or an environmental disaster. Epidemiologists, then, can pinpoint effective containment policies and monitor potential sources to prevent future outbreaks.

In longer outbreaks where a virus has enough time to mutate, contact tracing allows epidemiologists to study the viral evolution of an outbreak. During the COVID-19 pandemic, we saw many mutations of COVID-19 with different rates of mortality and different symptomologies, some of which did not respond well to current vaccines or treatments to which prior strains had responded. Early identification of new mutations and their geographic sources provides crucial information for quick response across countries and continents.

Let's consider an example of contact tracing in a new HIV outbreak. Let's say we are in a mid-sized city with many neighborhoods and subpopulations. Cases have popped up at several testing centers and hospitals over the past two months, and epidemiologists are concerned that the dominant strain in the area has mutated into a much deadlier strain of HIV, threatening the city. Two of the main ways that HIV spreads within a population are sexual contact and sharing syringes when using drugs intravenously; this hypothetical epidemic seems to be largely contained to young populations with high rates of drug use and risky sexual behaviors.

When a case is identified, healthcare providers collect information about others potentially exposed to HIV through sexual contact or syringe sharing such that the person or the provider can inform those exposed that they are at risk of HIV. Positive cases are identified and traced to build a network of exposure. Typically, metadata, such as strain genomics, demographic factors, geographic histories, and exposure routes, is collected and attached to this network to quickly identify the main sources of spreading and populations most at risk given identified case characteristics.

Determining causality is not always straightforward. Oftentimes, cases are missed when individuals do not present with illness or do not get tested regularly. This means cases identified later in an epidemic might be sources of infection. *Figure 12.2* shows a small piece of a contact tracing network, where cases identified later in the epidemic through contact tracing were infected earlier than the presenting individual (*Case #161*):

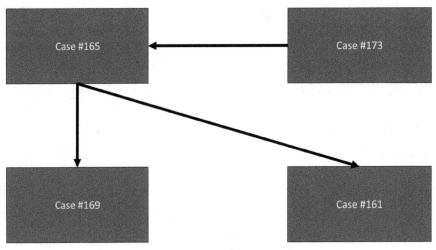

Figure 12.2 – A small subsection of a contact tracing network

In *Figure 12.2*, we see *Case #161* was identified (likely through a testing center or by seeking medical care). This individual's contacts revealed infection from *Case #165*, who was infected by *Case #173*. Contact tracing also revealed another individual infected from *Case #165* (*Case #169*). Without contact tracing, at least three other cases would have been missed, leading to treatment delays and potentially more exposures and cases. Generally, causality is established through strain analysis and information about the timing of exposures through sexual partners or syringe sharing. These factors are included in the case metadata.

Storing case metadata and causality information in a relational database would be very difficult, as the directionality of infection is critical for studying epidemic evolution over the population and for contacting potential exposures who have not yet been tested (as case contacts can overlap and often do in this type of epidemic). Graph databases allow for efficient storage and retrieval of this information for maximal efficiency of both contact tracing and public policy formation to protect those at risk of HIV exposure in this population. Early identification saves people from suffering and, potentially, death.

Island ecology study

Now that we know how graph databases improve analytics and outcomes for social science data, let's consider how graph databases can improve research in conservation. Suppose scientists have identified a new island in a remote area long isolated from mainland populations. *Figure 12.3* shows the hypothetical island:

Figure 12.3 – An illustration of an island long isolated from the mainland

On the island shown in *Figure 12.3*, we wish to document the flora and fauna that live on that island, as well as the island's food chains. We'll collect information on population sizes, predator-prey relationships, and potential threats to those populations and relationships (such as climate change). We may wish to link species into food webs to explore the sustainability of food sources given changes in temperature, weather, and sea levels on the island. We may also wish to divide the island into separate ecosystems, which may be isolated in valleys with high cliffs or connected to other ecosystems on the island through migration patterns or habitat overlap. Food webs and overlapping habitat/migration patterns create relationships between population data in our dataset, which naturally lives as a network. We can add metadata to the vertices in our network (species or habitats) or the edges connecting them (such as seasonal information about migration patterns that might connect species from two different habitats periodically).

In this conservation example, we'll likely have more than one dataset collected. We might have several different food webs from different geographies. We may also have habitat overlap datasets connecting geographies on the island that connect to our food webs, such as metadata on population size estimates, ecosystem health indicators, and seasonal patterns.

Connecting many datasets, some of which include hierarchical relationships, is quite tricky in relational databases; however, it is easy when using graph databases. Different types of edges can connect different vertices in collected networks to link each network and the metadata contained in each network collected. This makes querying and exploring potential relationships that are not explicitly defined (perhaps not collected due to time constraints or difficulty of terrain navigation) much easier for conservationists.

From this data and exploration of its graph database, it is possible to infer and define protected areas on the island in which many food webs connect or are isolated regions with large biodiversity. This protects species from human activities or climate change threats and allows researchers to focus limited funding on monitoring areas critical to the overall ecosystem. In addition, as further expeditions and conservation efforts collect new data, the graph database setup allows researchers to test hypothesized links between ecosystems or food webs that were not directly observed in the first pass expedition.

Now that we understand a bit about datasets that will benefit from graph database storage solutions, let's turn our attention to Neo4j, an open source graph database option.

Querying and modifying data in Neo4j

Before we start using Neo4j, we'll need to download and install the software on our machines. You can follow this link to reach the Neo4j Desktop download page: `https://neo4j.com/download/`.

You should see a page that looks like this, where you can download Neo4j by clicking on the **Download** button:

Figure 12.4 – Neo4j download site

After hitting the **Download** button in *Figure 12.4*, you'll be prompted to follow the installer instructions to complete the installation. When the installer finishes, you'll see an icon or will have launched Neo4j directly on your machine from the installation process. This should take you to a page that looks like this:

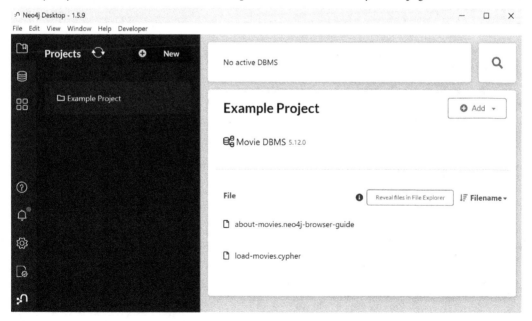

Figure 12.5 – The start page of Neo4j

Figure 12.5 shows the free Desktop version of Neo4j on a Windows machine.

> **Note**
> There is already an example project ready to explore or query.

We'll work on some basic queries with this data, but if you wish to explore this database prior to querying, click on **Movie DBMS** and hit the **Start** button to run the database on your machine; when you finish exploring, make sure to hit the **Stop** button to cache the database again.

The query language used by Neo4j is called **Cypher**. While it functions similarly to SQL, Cypher also contains graph theoretic operations to examine relationships and connectivity within the network. The `Match` command functions much like SQL's `SELECT` command, and operations such as `limit` and `as` also exist in Cypher. However, operations that retrieve directional relationships (edges) between objects in the database (vertices) also exist to query specific types of relationships that might exist. We'll see more of this in action as we explore the Movie DBMS (database management system) database on Neo4j's Desktop application.

Now, let's explore Neo4j.

Basic query example

When you open your Neo4j Desktop application, you'll see an `Example Project` folder. When you hover over the **Movie DBMS** label on the right-hand side of the screen, you'll see a **Start** button that launches the connection to this database (shown in *Figure 12.6*). Click on **Start**:

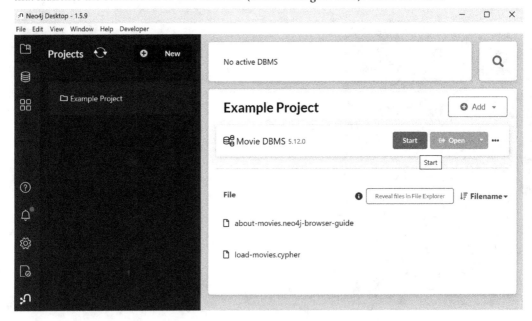

Figure 12.6 – The Start button to launch the Movie DBMS database

As shown in *Figure 12.6*, once you have launched the Movie DBMS database by clicking **Start**, you'll have a connection to this database. It may take a few minutes to launch this database. It is possible that you may not be familiar with Neo4j before reading this chapter, so we'll click on the about-movies.neo4j-browser-guide file option and run through a few examples of queries to become familiar with Cypher's syntax and results.

You'll see a page like *Figure 12.7* when you click on the about-movies.neo4j-browser-guide file option. The first page shows a basic query pulling data on an actor. Here, we'll change the actor to Brad Pitt and query the graph for instances of Brad Pitt. When you have the code changed to what is shown in *Figure 12.7*, hit the arrow to run the code. Note that you'll see the default tutorial below the code you write. We aren't showing this in our Neo4j query images but it will appear on your screen:

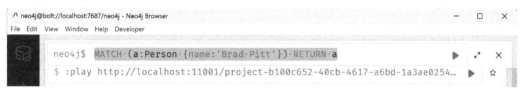

Figure 12.7 – A basic Cypher query that searches for Brad Pitt

This database does not contain any information about Brad Pitt. Running the suggested query on Tom Hanks does produce information in JSON form, as this data does not contain graph elements. You should see something like *Figure 12.8* when you query the database for Tom Hanks:

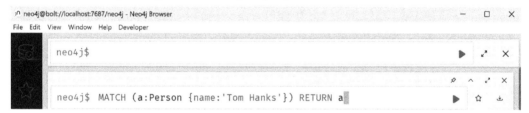

Figure 12.8 – The result of our query for Tom Hanks in Movie DBMS

Let's go to the next page of the guide and add an actor to the database. We'll add Brad Pitt, as our query came back empty initially. *Figure 12.9* shows how to add an actor to the database along with their birth year.

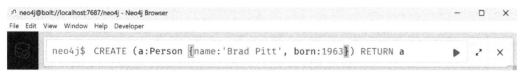

Figure 12.9 – Instructions to create an entry for Brad Pitt

When we run the code from *Figure 12.9*, Neo4j shows us a vertex with Brad Pitt's information, as seen in *Figure 12.10*:

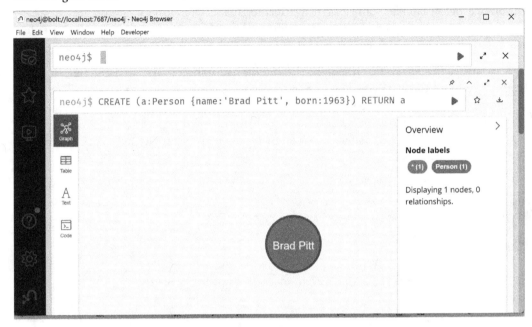

Figure 12.10 – The result of creating Brad Pitt's entry

We can add a movie associated with Brad Pitt, in this case, Seven Years in Tibet, as shown in *Figure 12.11*'s code in the top line:

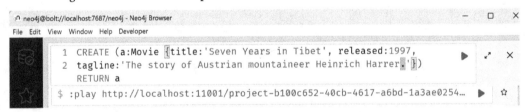

Figure 12.11 – Cypher code to add a movie associated with Brad Pitt

When we run this Cypher code, we see another graph database vertex appear, as shown in *Figure 12.12*:

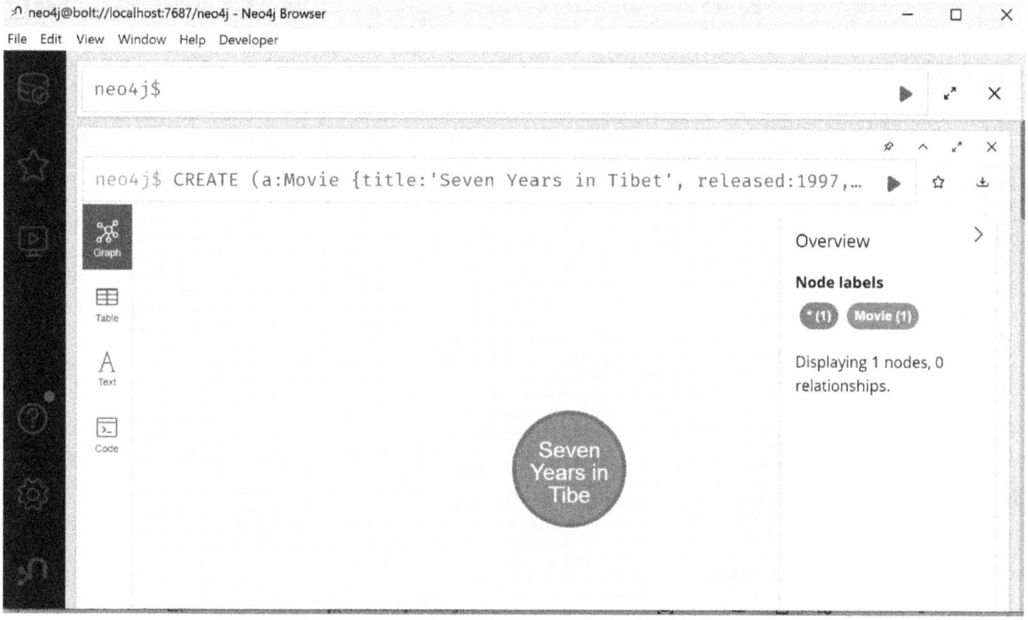

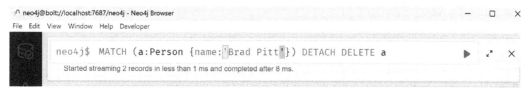

Figure 12.12 – The result of running the Cypher query from Figure 12.11

Just as we can add to our graph database, we can also delete individual records from the database. Let's delete the records we just created for `Brad Pitt`. *Figure 12.13* shows the deletion of the record for `Brad Pitt`. I urge you to try deleting `Seven Years in Tibet` before moving on to the next task in this chapter:

Figure 12.13 – Cypher code to delete the record we created for Brad Pitt

We can also create or update a record in the database using the MERGE statement with a CREATE and a MATCH statement to either create a new record or update an existing record (depending on what exists in the database). We'll add Brad Pitt's information back into the database using the code shown in *Figure 12.14*:

```
1  MERGE (a:Person {name:'Brad Pitt'})
2  ON CREATE SET a.born = 1989
3  ON MATCH SET a.stars = COALESCE(a.stars, 0) + 1
4  RETURN a
```

Figure 12.14 – An updated example in Cypher code

When we run the code in *Figure 12.14*, we see *Figure 12.9* appear again, where Brad Pitt's information has been added to the database. Now that we know some basic Cypher commands, we can move on to more complicated graph database operations in the next section.

More complicated query examples

Now that we know how to execute basic queries, we can move on to more complicated operations in Neo4j, such as adding relationships between items. Let's connect actors and movies with a Cypher query connecting Brad Pitt to *Seven Years in Tibet* through his character, Heinrich Harrer.

> **Note**
> You'll need to add both the movie and actor back into our database using the code in the *Basic query example* section before running this query. This query is shown in *Figure 12.15*.

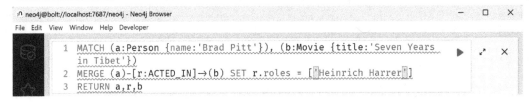

```
1  MATCH (a:Person {name:'Brad Pitt'}), (b:Movie {title:'Seven Years
   in Tibet'})
2  MERGE (a)-[r:ACTED_IN]→(b) SET r.roles = ['Heinrich Harrer']
3  RETURN a,r,b
```

Figure 12.15 – A Cypher query to connect Brad Pitt to his character, Heinrich Harrer, in Seven Years in Tibet

Running this query returns information about the actor, the movie, and the relationship we just created between the two datasets, as shown in *Figure 12.16*:

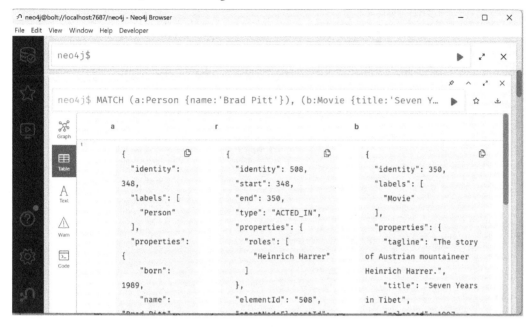

Figure 12.16 – The result of Figure 12.15's Cypher query to connect Brad Pitt to Seven Years in Tibet

The next page of the Neo4j Cypher guide shows an alternative way to create this relationship. You are encouraged to modify that code to add another actor and movie as an exercise. Let's turn our attention to Cypher's WHERE clauses, which function much as they do in SQL.

We'll search for persons whose names start with the first name Brad, which should pull up Brad Pitt's entry, as well as any other actors whose first name is Brad. *Figure 12.17* shows this query:

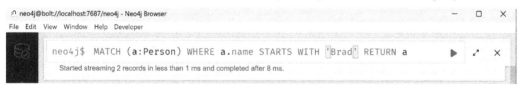

Figure 12.17 – An example of a WHERE query in Cypher

When we run the Cypher code in *Figure 12.17*, we should see two instances of Brad Pitt showing up in our database, as shown in *Figure 12.18*:

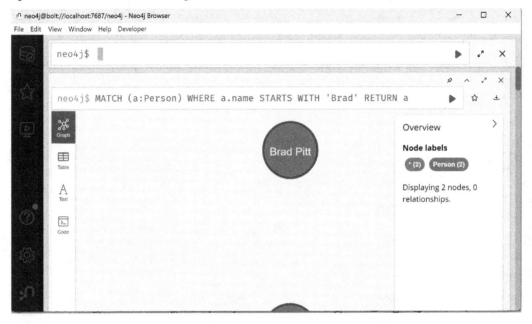

Figure 12.18 – Results from Figure 12.17's WHERE Cypher query

We can also match based on relationships, such as finding movies associated with an actor. Let's try a Cypher query on relationships to find all of the movies associated with Brad Pitt (here, just Seven Years in Tibet, as we added this relationship). *Figure 12.19* shows the query we'll use to query a relationship:

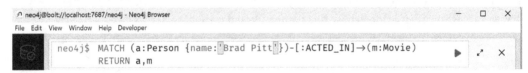

Figure 12.19 – A Cypher query on graph relationships between people and movies

When we run the Cypher query shown in *Figure 12.19*, we see the connections between our instances of Brad Pitt and Seven Years in Tibet shown in graph form (*Figure 12.20*):

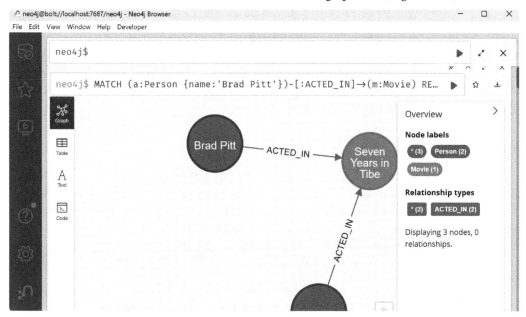

Figure 12.20 – The results of Figure 12.19's graph relationship query

Figure 12.20 shows the essential strengths of a graph database. We can see relationships that connect different items across tables (here, the Person table and the Movie table), as well as the nature of that relationship. It's easy to see an actor connected to a movie, and should more movies exist, we'd have an easy visualization of our query results from this graph-based query in Neo4j.

While this movie database contains basic tables and relationships, we can imagine our examples in the *What can you represent in a graph database?* section containing much more complicated relationships between many more tables. Visualizing query results in graph form across tables and items in tables provides easy-to-understand summaries of the data and relationships of interest in a query. Herein lies the power of graph databases. Queries tend to be quick, even for queries with many subqueries or when searching very large tables and their relations, and visualizations of results allow users to explore relationships in depth across tables (or provide succinct visuals for a report).

If you wish to explore Neo4j further, Neo4j's AuraDB allows users to connect to the database through Python by establishing an account and connecting to Neo4j with an API in Python using the neo4j package and GraphDatabase.driver() function to create the connection.

> **Note**
> For any large databases or databases you wish to permanently save, you'll need a paid account.

Once the database is created, you can query the Neo4j database through Python's `neo4j driver.execute_query()` statement and import the data as a network in igraph to analyze the results. This provides a powerful combination of graph storage and graph analysis for your analytics projects. If you are interested, you are encouraged to explore Neo4j further and create your own connected projects between igraph and Neo4j's AuraDB accounts.

Summary

In this chapter, we examined cases where graph databases are advantageous, familiarized ourselves with an open source graph database called Neo4j, and learned a bit about the query language of Neo4j, called Cypher. We created, deleted, and modified records in a Neo4j movie database. We explored the advantages of querying graph databases and the unique query result visualizations possible with graph databases. If you are interested, I encourage you to consult Cypher and Neo4j resources to dive deeper into what is possible with graph databases.

In the next chapter, we'll be putting together all of the skills we've learned in the book so far to tackle a real-world problem of predicting Ebola outbreak severity over time and geography across regions of the Democratic Republic of Congo.

References

Adler, M. W., and Johnson, A. M. (1988). *Contact tracing for HIV infection. British Medical Journal (Clinical research ed.), 296*(6634), 1420.

Angles, R., Arenas, M., Barceló, P., Hogan, A., Reutter, J., and Vrgoč, D. (2017). *Foundations of modern query languages for graph databases. ACM Computing Surveys (CSUR), 50*(5), 1–40.

Beas-Luna, R., Novak, M., Carr, M. H., Tinker, M. T., Black, A., Caselle, J. E., ... and Iles, A. (2014). *An online database for informing ecological network models*: `http://kelpforest.ucsc.edu`. *PloS one, 9*(10), e109356.

Hyman, J. M., Li, J., and Stanley, E. A. (2003). *Modeling the impact of random screening and contact tracing in reducing the spread of HIV. Mathematical biosciences, 181*(1), 17–54.

Mueller, W., Rudowicz-Nawrocka, J., Otrzasek, J., Idziaszek, P., and Weres, J. (2016). *Spatial data and graph databases for identifying relations among members of cattle herd. International Multidisciplinary Scientific GeoConference: SGEM, 1*, 835–841.

Pasquale, D. K., Doherty, I. A., Leone, P. A., Dennis, A. M., Samoff, E., Jones, C. S., ... and Miller, W. C. (2021). *Lost and found: applying network analysis to public health contact tracing for HIV. Applied network science*, 6, 1–16.

Pokorný, J. (2015). *Graph databases: their power and limitations.* In *Computer Information Systems and Industrial Management: 14th IFIP TC 8 International Conference, CISIM 2015, Warsaw, Poland, September 24-26, 2015, Proceedings 14* (pp. 58-69). Springer International Publishing.

Robinson, I., Webber, J., and Eifrem, E. (2015). *Graph databases: new opportunities for connected data.* O'Reilly Media, Inc..

Shannon, P., Markiel, A., Ozier, O., Baliga, N. S., Wang, J. T., Ramage, D., ... and Ideker, T. (2003). *Cytoscape: a software environment for integrated models of biomolecular interaction networks. Genome research*, 13 (11), 2498–2504.

Webber, J. (2012, October). *A programmatic introduction to neo4j.* In *Proceedings of the 3rd annual conference on Systems, programming, and applications: software for humanity* (pp. 217–218).

13

Putting It All Together

In the previous chapters, we explored different data problems amenable to network science solutions, learned many network centrality metrics, and applied **machine learning** (**ML**) models to network science datasets. In this chapter, we'll put together what we have learned so far to assess epidemic risk on a spatiotemporal dataset assessed at each time point for centrality metrics and number of Ebola cases. Our example will include network and other data collected on a geographic region of the Democratic Republic of Congo that was impacted by a large Ebola outbreak in 2018-2020 and several smaller outbreaks since 2020. We'll also introduce a statistical model used to model time series data much like a regression model.

Specifically, we will cover the following topics in this chapter:

- Introduction to the problem
- Introduction to **generalized estimating equations** (**GEEs**)
- Data transformation
- Data modeling

By the end of this chapter, you'll understand how to solve a problem end-to-end with network science tools and statistical models. You'll be able to run time series models in Python and interpret the results mathematically. Finally, you'll end this chapter equipped to apply what you have learned in this book to data science problems that you face in your work.

Technical requirements

The code for the practical examples presented in this chapter can be found here: `https://github.com/PacktPublishing/Modern-Graph-Theory-Algorithms-with-Python`

Introduction to the problem

The Ebola virus is a **filovirus**, a single-stranded ribonucleic acid (RNA) virus enveloped with a lipid layer. It was first identified in 1976 and is **zoonotic**—spreading from infected animal hosts to human populations. The animal vector from which Ebola spreads to humans is currently unknown. However, the virus is spread from person to person through direct contact with bodily fluids, placing frontline healthcare workers and family caregivers at high risk of both contracting and spreading the virus.

The Ebola virus inhibits immune response, inhibits clotting, promotes renal and electrolyte dysfunction, and promotes acidosis. All of these contribute to hypovolemia, which triggers organ failure and death in many cases. It is endemic to several areas of Africa, including the Congo Basin, where the majority of outbreaks have originated. Case fatality rates are typically in the 40-50% range in known outbreaks, typically due to a lack of supportive care such as transfusions and life support in rural regions that often experience the initial cases.

During outbreaks, personal protective gear is critical for avoiding spread to and by healthcare providers. However, in many rural areas, lack of medical supplies limits healthcare providers' ability to protect themselves, contributing to early disease spread. In addition, isolation of cases is necessary to avoid infecting other patients seeking care at a clinic. *Figure 13.1* shows a doctor in personal protective gear at a clinic set up to isolate Ebola patients:

Figure 13.1 – An illustration of a doctor in personal protective
gear at a makeshift clinic to treat Ebola patients

In December 2013, the worst outbreak of the Ebola virus began in Meliandou in Guéckédou Prefecture, Guinea—a small town from which the virus spread through Guinea and its neighbors, Sierra Leone and Liberia. Eventually, cases spread to Italy, Mali, Nigeria, Senegal, Spain, the United Kingdom, and the United States, with secondary cases reported in several of these countries (mainly healthcare workers). By the end of the outbreak, nearly 30,000 people were infected, and over 10,000 died. The virus mutated multiple times by the end of the outbreak in 2016, and researchers developed a vaccine to extinguish the epidemic and limit its spread around the world and within large cities such as Conakry, Guinea. The reproduction number ranged from 1.20 in Liberia to 2.02 in Sierra Leone, contributing to the rapid spread and high infection rate in the three main countries impacted.

The 2018 Ebola virus outbreak in the Democratic Republic of Congo impacted the provinces of Ituri, North Kivu, and South Kivu, with cases concentrated in Beni, Butembo, Katwa, Kalunguta, Mabalako, and Mandima. Over 3,500 confirmed and suspected cases were identified by 2020, with over 2,200 deaths despite prompt intervention from local and international public health workers and the availability of a vaccine developed during the 2014 West African Ebola outbreak. From late 2018 to mid-2019, the reproduction number remained over 1, indicating sustained infection potential. In fact, most cases did occur during this time frame.

Prompt intervention limited the potential number of infected people to roughly 16,000—rather than 16 million people living in the impacted provinces—due to the isolation of towns and prompt interventions by public health workers. Border surveillance measures limited the spread of Ebola to neighboring countries such as Burundi and Rwanda. However, distrust of international aid workers, the ongoing conflict with the M23 rebel group, and violence spilling into treatment areas complicated the epidemic response by local and international aid groups. Better coordination of response and better handling of security threats could have prevented the spread and led to better survival rates.

While this chapter will focus on modeling networks that facilitated the spread of Ebola across regions of the Democratic Republic of Congo, the methods we'll use can be extended to working with any type of **partial differential equation** (**PDE**) model related to the spread of networks. For instance, readers interested in ecology and conservation might consider a network science approach to modeling the geographic spread of toxins that impact leafy sea dragons, Bahamian saw sharks or sulcata tortoise populations; they may also want to understand evasion or delays in predator-prey relationships. Social scientists might consider the spread of political ideologies across countries in a region or models where effects of behavior change at a population level in response to social media campaigns may involve delays, where change happens over longer periods of time or needs to reach a threshold before effects are seen. Biomedical researchers might consider chemical gradient models within the context of cell differentiation and morphogenesis, the spreading of seizures through the cerebral cortex, or even the dynamics of tumor growth across regions of the brain. Farmers may wish to consider spatial models that limit the spread of locusts or other pests that endanger crops with limited barrier resources.

For now, let's return to our Ebola virus example and dive deeper into the context of the outbreak.

Ebola spread in the Democratic Republic of Congo – 2018-2020 outbreak

Despite a rapid response from public health workers, the 2018 Ebola outbreak included waves of exponential growth. Many impacted areas included cities with populations over 100,000, and militant attacks on healthcare facilities prior to the outbreak and since the outbreak have threatened medical infrastructure and prompt care for patients. Due to the ongoing conflict, many at-risk populations in rural areas outside major cities hesitated to receive vaccines or follow public health advisories. Misinformation was abundant, delaying treatment for some patients and allowing Ebola to spread to caregivers. All these factors facilitated the spread of Ebola from city to city within the impacted regions.

However, much of Ituri, North Kivu, and South Kivu includes regions of dense rainforests, wildlife preserves, and sparse populations, limiting travel between cities that typically fuels epidemic spread across regions (see *Figure 13.2*). In fact, much of the spread occurred across larger cities connected by transportation. This likely ameliorated the impact of this Ebola epidemic compared to the West African 2014 outbreak that included cities with many transportation routes:

Figure 13.2 – An illustration of a protected area of the Congo Basin

Unfortunately, due to cities nested in densely forested regions where Ebola's animal vectors live, the geography of the northeastern Democratic Republic of Congo increases the likelihood of outbreaks in the region. This increases the likelihood of future Ebola outbreaks in the region. Thus, it is important to study some of the factors that facilitate epidemic spread and vulnerability over time, such that quick and effective responses reach the region as soon as another outbreak is identified. This limits the number of people exposed and potentially lowers mortality rates as more people are vaccinated and treated.

Geography and logistics

As we have seen in the prior subsections, regional geography and the logistics of transportation between cities play a large role in the potential spread of an epidemic. Within an epidemic, an isolated area with little or no population mixing with other regions suggests a contained epidemic. However, a transportation hub (such as a city with many international flights per day) poses a much greater risk to public health during an outbreak. In fact, this is exactly what quickly spread COVID-19; short incubation periods coupled with many travelers from international hubs traveling to other international hubs led to an explosion of cases in new countries and continents until borders were shut down to stem the flow of disease.

In our analysis, we'll consider a few of the regions most impacted by the 2018 Ebola outbreak, including network connectivity metrics based on transportation routes between regions, number of cases per year, number of violent incidents (which may displace people regionally), and number of disaster incidents (again, a regional displacement risk) from 2017-2021. We hypothesize that network metrics derived from regional connectivity will impact the number of Ebola cases during the outbreak period.

Before we begin constructing our dataset, let's first establish the necessary tools for our analysis—specifically, longitudinal regression modeling.

Introduction to GEEs

Regression models come with many assumptions that need to be relaxed for many types of real-world problems. For instance, linear regression assumes a normally distributed outcome. In many problems, we may wish to work with binary outcome data (yes/no, survived/died, recurred/did not recur…), count data (number of events in a period of time, such as this chapter's outcome), or failure rates for a manufactured product (likelihood of failure outcome).

One of the most common outcome distributions comes from the binomial distribution, in which binary data is collected across a population. For example, we may have a sample of patients in a glioblastoma study, where we compare 6-month survival rates of patients across different treatment groups. Each patient will either survive or die (binary outcome); aggregated by group, these outcomes form binomial distributions, which can be compared statistically to determine if an optimal treatment exists for glioblastoma patients.

Generalized linear regression extends linear regression to outcomes such as the aforementioned ones by using a link function to transform the outcome to that of a normal distribution using a geometry-rooted map; most types of outcomes fit into this framework, as they belong to an outcome family called the **exponential family**. Distributions in the exponential family can be mapped to each other using tools from geometry.

However, generalized linear regression comes with many assumptions common to linear regression: linear independence of variables, linear relationships between predictor variables and the outcome, and constancy of errors across levels of predictor variables (technically called homoscedasticity), among others. Some of these assumptions—namely, independence of predictor variables—do not hold for time series or spatial data. Predictor variables measured at one time point likely depend on values of those variables measured at prior time points. In our example, the ability of public health workers and doctors to identify an outbreak early enough to stop it plays a large role in later case rates and resource needs (shown in *Figure 13.3*):

Figure 13.3 – An illustration of doctors and epidemiologists tracking the source of an outbreak

GEEs are one solution that mathematically transforms generalized linear models to data involving multiple time points or clustering within the dataset. Essentially, we can model dependencies across time (or clusters) using a matrix of dependency correlations. GEEs retain a link function, allowing them to model outcomes within the exponential family, such as our number of yearly Ebola cases across regions.

One of the advantages of this approach compared to other generalized linear model extensions for time series or clustered data is that GEEs are easier to compute. Many other extensions, such as generalized linear mixed models, require multilevel modeling, with dependencies across levels. This creates additional parameters and dependencies within the model (which create conditional distributions). GEEs do not involve conditional distribution calculations. This allows us to save on computational costs, even with simple time series or clustering components in our model.

For readers who have a more advanced mathematical background, we'll spend a bit of time diving into the mathematics of generalized linear models and GEEs in the next section. Readers less comfortable mathematically can skip the mathematical formalization and go directly into the data construction part of the *Our problem and GEE formulation* section. However, it's helpful to know the mathematics behind regression equations when using them, and those of you interested in using GEEs are encouraged to look through the references at the end of the chapter for more detailed mathematical explanations.

Mathematics of GEEs

In **linear regression**, where the outcome is normally distributed and observations and predictors are independent of each other, we estimate the effect sizes of each predictor through estimations of matrix algebra, where the outcomes are known for each observation (represented as a **vector**) and the predictors are known for each observation (represented as a matrix of vectors associated with each predictor). The outcome vector equals the effect-size vector multiplied by the predictor matrix plus an error term to capture random errors in measurement. In practice, the problem is difficult to solve with matrix operations and an algorithm to estimate the effect sizes iteratively is used to solve it.

Generalized linear regression extends this methodology by introducing a link function applied to the outcome vector to transform the equation to a linear form, for which effect size coefficients are found by applying an algorithm. However, to rescale those effect sizes so that they match the scale of the original outcome prior to the application of the link function, we reverse the operations of the link function to scale the effect size appropriately. In logistic regression, for instance, a common link function involves log transformation. Exponentiating the effect sizes rescales those effect sizes so that they match the outcome's scale rather than the log of the outcome.

In practice, linear and generalized linear regression models are fit using **maximum likelihood estimation** (**MLE**), an algorithm that iteratively adjusts estimates of the effect-size vector. This algorithm fits according to the assumed probability distribution of the outcome and the observed distribution of the predictors by iterative resampling. This allows for statistical inference on the estimation of effect sizes.

With respect to GEEs, the clustered nature of the outcomes and predictors by individuals in our sample over time introduces a correlation structure for each individual in our sample, as well as sample-based effect sizes for predictors based on outcomes. We define a best guess as to the structure of the correlation matrix within individuals, called the **autoregressive structure**, which can be estimated from the data, left unspecified, or defined as a specific structure tying observations together across time or space. In practice, leaving this unspecified or estimated from the data is best if the correlation structure is not known. After defining the autoregressive structure, we estimate the covariance matrix to correct for any assumption errors within the autoregression specifications.

An algorithm then estimates the effect sizes within and across individuals. Generally, algorithms used to fit the model rely on computing the derivatives of the GEE's linear algebra formulation rather than the computation of maximum likelihood estimators; derivative-based methods often offer computational advantages over the MLE methodology. In addition, derivative-based methods yield model output similar to a generalized linear regression model, though not all overall model fit statistics can be estimated.

Those of you who are interested in the linear algebra formulation of generalized linear regression and GEEs can consult the references at the end of this chapter for detailed explanations of these models and the estimators used to find predictor effect sizes. The Nelder (1972) text overviews the theory of generalized linear models and their computation very nicely. The Hanley et al. (2003) text provides both a linear algebra overview of GEEs and a simple case study where estimates can be calculated easily by readers. It's recommended that those of you who wish to understand the mathematics behind GEEs start with the mathematics and computational methodology of generalized linear models before tackling models derived from them, such as GEEs.

For now, let's turn back to the construction of our dataset.

Our problem and GEE formulation

In our formulation of Ebola risk over time and geography in the Democratic Republic of Congo, we'll look at the number of yearly Ebola cases in one of the following cities as our outcome:

- Mandima
- Mabalako
- Kangulata
- Katwa
- Butembo
- Bunia
- Beni

We'll cluster these cities by province, including Mandima and Bunia in Ituri and Mandima, Mabalako, Kangulata, Katwa, Butembo, and Beni in North Kivu. Next, we'll construct a network of transportation options between cities and obtain betweenness and degree of centrality for each city. This will provide us with an estimate of bridgeness and hubness within each year's network.

We'll also include other statistics, including the number of violent incidents that year involving our cities and the number of natural disasters impacting our cities.

For a number of Ebola cases, we'll consolidate the `ebola-epidemic-health-zone-figures.csv` **Humanitarian Data Exchange (HDX)** dataset (`https://data.humdata.org/dataset/ebola-cases-and-deaths-drc-north-kivu`) to include the cities of interest for our analysis. To obtain the necessary transportation, we'll search the web for feasible routes between cities (biking, public transportation, or driving). Given that this is current data, we will also search for changes within the period of 2017-2021, including COVID-19 restrictions. We'll use **ReliefWeb** to find incidents mentioning natural disasters (`https://reliefweb.int/disasters?list=Democratic%20Republic%20of%20the%20Congo%20Disasters&advanced-search=%28C75%29`).

Consolidating another HDX dataset will yield information about violent incidents in each area (`https://data.humdata.org/dataset/ucdp-data-for-democratic-republic-of-the-congo`). While this may not include all information for a region in our time period, we should capture major conflicts or disasters.

Now that we have our data sources, let's explore the data and perform the needed transformations to coax the data into a format that will work as input to our GEE model.

Data transformation

As we assembled our dataset, we noticed a few caveats in data quality worth mentioning. Very few natural disasters hit the area during our time period of interest, and these were spread across a wide geography. It's unlikely that this factor will impact Ebola spread, but we will keep this variable in our dataset.

The source for violent incidents appears to be missing information, as known attacks on aid workers and Ebola treatment sites in Katwa are not included in the data. This suggests an incomplete data source that may not capture violent incidents at the level needed for a real-world analysis of factors influencing Ebola spread. However, for purposes of demonstrating this method, it captures enough to be potentially interesting in the analysis. In projects such as this, data quality can be questionable, as good sources are hard to find in much of the developing world.

The search for transportation routes was also questionable, yielding some insight into direct routes but not in as much detail as would be ideal. We considered any route between city pairs to be viable if travel would take less than 10 hours by car, bus, or foot travel. Most cities yielded useable data; however, some cities did not show viable routes, yielding isolated vertices in our travel network. No major differences in routing existed across the years (no projects listed to create new roads or bus routes), so our network properties will not change over the years.

The Ebola case source data likely showed some errors, as cumulative cases were sometimes lower over time for certain locations. We erred on the side of retaining positive or zero case totals per year, given that we hope to model our outcome as count data. Cases can be difficult to determine in such outbreaks, as lab confirmation is difficult and often time-consuming when sent to another city for analysis.

Thus, our data collection for this project highlights some of the real-world challenges of building datasets from scratch, particularly in areas where data collection may be incomplete or nonexistent with respect to open source sites. It is often advantageous to examine many potential sources and either merge insights or choose the best source. In our case, we chose the most complete sources to obtain our dataset.

Python wrangling

From our sources, we've assembled a .csv file containing year, city, region, conflict, disaster, and Ebola case data. However, we'll need to compute **betweenness centrality** and **degree centrality** from our transportation network and add that to our dataset to capture transportation network properties for each city.

Let's get started by computing the centrality metrics for each city in our transportation dataset so that we can append these values to our .csv file. We'll compute the metrics using Script 13.1:

```
#install igraph and pycairo
!pip install igraph
!pip install pycairo


#import igraph, pandas, and numpy
import igraph as ig
from igraph import Graph
import numpy as np
import pandas as pd
import os


#import adjacency matrix
File ="C:/users/njfar/OneDrive/Desktop/DRC_Transport.csv"
pwd = os.getcwd()
os.chdir(os.path.dirname(File))
mydata =
    pd.read_csv(os.path.basename(File),encoding='latin1',header=None)


#create and plot graph
g_transport=Graph.Adjacency(mydata)
ig.plot(
    g_transport,bbox=(200,200),
    vertex_label = ["Beni","Bunia","Butembo","Kalunguta","Katwa",
        "Mabalako","Mandima"]
)


#calculate betweenness and degree centrality
bet=Graph.betweenness(g_transport)
deg=Graph.degree(g_transport)


#create vector to add to dataset
degree=deg+deg+deg+deg+deg
betweenness=bet+bet+bet+bet+bet
```

This script should yield a plot showing two isolated vertices and several vertices connected by weighted paths with values of 1, 2, or 3, as shown in *Figure 13.4*:

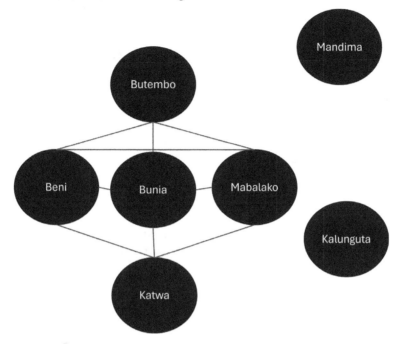

Figure 13.4 – A plot of the Ebola network city

From *Figure 13.4*, we can see that Kalunguta and Mandima are not connected to other cities, while other cities seem to be connected quite densely into a high-traffic area. For a more comprehensive examination of travel, we might also consider the volume of travel between areas; within our context, this data does not seem to exist.

Now that we have our vectors of betweenness and degree centrality, which capture the bridgeness and hubness of our cities, we can attach these values to the dataset we've created to ready the dataset for analysis with a GEE.

GEE input

Now that we've derived our network metrics, let's import our collected dataset and attach the network metrics as additional columns by adding to Script 13.1:

```
#import Ebola dataset
File ="C:/users/njfar/OneDrive/Desktop/Ebola_Data.csv"
pwd = os.getcwd()
os.chdir(os.path.dirname(File))
```

```
data = pd.read_csv(os.path.basename(File),encoding='latin1')
#add degree and betweenness to Ebola dataset
degree=np.array(degree)
betweenness=np.array(betweenness)
data['Betweenness']=betweenness
data['Degree']=degree
```

Now that we've assembled our full dataset, we're ready to analyze it with our GEE model.

Data modeling

For our model, we'll assume our outcome has a **Poisson distribution**. While our dataset doesn't have a lot of zero values, it may be worth considering a zero-inflated Poisson or negative binomial distribution rather than a Poisson distribution. However, for simplicity of modeling, we will set our family to a Poisson distribution.

We'll use a data-derived autocorrelation structure, as a best guess for correlations over time within towns isn't readily apparent. In practice, it's usually better to derive this structure from data than make a guess when the structure isn't known well beforehand.

Let's get started modeling this GEE in Python.

Running the GEE in Python

Let's first import our GEE model and define the distribution family and covariance structure with `Script 13.2`:

```
#load packages
import statsmodels.api as sm
import statsmodels.formula.api as smf

#define GEE parameters
family=sm.families.Poisson() #count data

#data-derived covariance structure
cov_str=sm.cov_struct.Exchangeable()
```

Now that we have our model defined, we can fit the model to our data, including all of our relevant predictors by adding to `Script 13.2`:

```
#create GEE model predicting Ebola cases
Model = smf.gee(
    "Ebola_Cases~Province+ViolentIncidents+Disasters+Degree+Betweenness",
    "Town",
    data,
```

```
    cov_struct=cov_str,
    family=family
)
results=model.fit()
```

This model should fit quickly given the small size of the dataset. Larger models may take a while to run, particularly those with large numbers of clusters and many time points considered. However, this model will run more quickly than other longitudinal models such as generalized linear mixed models, as it does not have a hierarchical structure or likelihood function derivations.

We can now examine our model's parameters and fit statistics to see what is predictive of Ebola cases within our data by adding to `Script 13.2`:

```
#examine resulting model
results.summary()
```

This should give a summary similar to *Table 13.1*:

GEE Regression Results			
Dep. Variable:	Ebola_Cases	No. Observations:	35
Model:	GEE	No. clusters:	7
Method:	Generalized	Min. cluster size:	5
	Estimating Equations	Max. cluster size:	5
Family:	Poisson	Mean cluster size:	5.0
Dependence structure:	Exchangeable	Num. iterations:	11
Date:	Thu, 21 Dec 2023	Scale:	1.000
Covariance type:	robust	Time:	10:18:47

	coef	std err	z	P>|z|	[0.025	0.975]
Intercept	3.8291	0.716	5.345	0.000	2.425	5.233
Province[T.North_Kivu]	1.1145	0.699	1.593	0.111	-0.256	2.485
Violent_Incidents	0.0312	0.022	1.415	0.157	-0.012	0.074
Disasters	-3.3828	0.323	-10.467	0.000	-4.016	-2.749
Degree	-0.0098	0.093	-0.105	0.917	-0.192	0.173
Betweenness	-0.0806	0.975	-0.083	0.934	-1.991	1.830

Skew:	1.7851	Kurtosis:	3.7263
Centered skew:	1.3387	Centered kurtosis:	2.2241

Table 13.1 – GEE model fit statistics

Table 13.1 shows that most of our predictors do not show a relationship with the number of Ebola cases in each town. Disasters do seem to have a negative relationship (z-value of -10.467, which is significant at $p<0.001$). This might relate to relevant resources positioned in an area to respond to a disaster that can be repurposed for response to the Ebola outbreak. Note that violent incidents reach a level of near significance statistically (z-value of 1.415, which is a p-value of 0.157), showing a potentially positive relationship with the number of cases. In a larger sample and with better data on violent incidents, it is likely that violent incidents would turn out to be predictive of Ebola cases, as we had hypothesized. The location of North Kivu appears to have another nearly significant relationship (p-value of 0.111), suggesting that it is an important location that requires monitoring. The other potential predictors have z-values near 0, suggesting that they are not important and likely would not play a role even with a larger sample size.

We do not see either of our network metrics as a significant predictor of Ebola cases. It is likely that transportation does not play a large role in outbreak case volumes; given the relative isolation of some areas, this makes logical sense. Yearly travel volumes between cities, if they could be obtained, might be more predictive of Ebola cases over time. However, to our knowledge, this does not exist for the towns considered in our analysis.

Summary

In this chapter, we tied together many of the tools we examined in previous chapters to tackle a major problem in public health: Ebola outbreaks. We constructed and analyzed properties of spatiotemporal data in a region recently impacted by an Ebola outbreak. We then fed this information into a longitudinal statistical model called a GEE to understand which factors may have contributed to the outbreak's spread over time and geography using Python.

In this book so far, we have overviewed many tools and now applied them on a final project. Next, we'll consider nascent tools in network science that offer promise for future projects, including hypergraphs and quantum network science algorithms.

References

Ballinger, G. A. (2004). Using generalized estimating equations for longitudinal data analysis. *Organizational research methods, 7(2), 127-150.*

Bolker, B. M., Brooks, M. E., Clark, C. J., Geange, S. W., Poulsen, J. R., Stevens, M. H. H., & White, J. S. S. (2009). Generalized linear mixed models: a practical guide for ecology and evolution. *Trends in ecology & evolution, 24(3), 127-135.*

Breslow, N. E. (1996). Generalized linear models: checking assumptions and strengthening conclusions. *Statistica applicata, 8(1), 23-41.*

Gatherer, D. (2014). The 2014 Ebola virus disease outbreak in West Africa. *Journal of General Virology, 95(8), 1619-1624.*

Groseth, A., Feldmann, H., & Strong, J. E. (2007). The ecology of Ebola virus. *Trends in microbiology, 15(9), 408-416.*

Guetiya Wadoum, R. E., Sevalie, S., Minutolo, A., Clarke, A., Russo, G., Colizzi, V., ... & Montesano, C. (2021). The 2018–2020 ebola outbreak in the Democratic Republic of Congo: A better response had been achieved through inter-state coordination in Africa. *Risk management and healthcare policy, 4923-4930.*

Hanley, J. A., Negassa, A., Edwardes, M. D. D., & Forrester, J. E. (2003). Statistical analysis of correlated data using generalized estimating equations: an orientation. *American journal of epidemiology, 157(4), 364-375.*

Holmes, E. C., Dudas, G., Rambaut, A., & Andersen, K. G. (2016). The evolution of Ebola virus: Insights from the 2013–2016 epidemic. *Nature, 538(7624), 193-200.*

Jacob, S. T., Crozier, I., Fischer, W. A., Hewlett, A., Kraft, C. S., Vega, M. A. D. L., ... & Kuhn, J. H. (2020). Ebola virus disease. *Nature reviews Disease primers, 6(1), 13.*

Liang, K. Y., & Zeger, S. L. (1986). Longitudinal data analysis using generalized linear models. *Biometrika, 73(1), 13-22.*

McKinney, W., Perktold, J., & Seabold, S. (2011). Time Series Analysis in Python with statsmodels. *Jarrodmillman Com, 96-102.*

Nelder, J. A., & Wedderburn, R. W. (1972). Generalized linear models. *Journal of the Royal Statistical Society Series A: Statistics in Society, 135(3), 370-384.*

Vossler, H., Akilimali, P., Pan, Y., KhudaBukhsh, W. R., Kenah, E., & Rempała, G. A. (2022). Analysis of individual-level data from 2018–2020 Ebola outbreak in Democratic Republic of the Congo. *Scientific Reports, 12(1), 5534.*

WHO Ebola Response Team. (2014). Ebola virus disease in West Africa—the first 9 months of the epidemic and forward projections. *New England Journal of Medicine, 371(16), 1481-1495.*

14
New Frontiers

In the previous chapters, we overviewed many of the tools and applications of network science within analytics projects. In this chapter, we'll look ahead toward the newer tools being developed that have many promising applications within network science, including quantum graph algorithms, deep learning/large language model architecture optimization, and multilevel graphs that are useful for organizing metadata and understanding genetics data.

While the prior chapters included coded examples, this chapter will focus on ideas and the possibilities for development in the future. Network science is an evolving field, and it's likely that tools we can't even imagine right now will be commonplace in the next decade. Let's dive into some of the newer applications and see how network science continues to contribute to knowledge in many different fields.

Specifically, we will cover the following topics:

- Quantum network science algorithms
- Neural network architectures as graphs
- Hierarchical networks
- Hypergraphs

Quantum network science algorithms

One new and promising avenue for network science algorithm development is quantum computing. **Quantum computing** leverages many of the advantageous properties of physics at the quantum level to improve computing power and tackle difficult problems. While there are many flavors of quantum computing, we'll focus on **qubit** systems, where bits are replaced with their quantum version.

Qubits offer many advantages over bits within computing frameworks. **Superposition** is a property in quantum physics that allows particles, such as qubits, to exist in multiple states at the same time. Thus, whereas bits must exist in a 0 or 1 configuration, qubits can exist as a 1 and a 0 simultaneously until the qubit is measured, collapsing to the usual state of a bit. This allows for massively parallel searches for solutions.

In addition to being able to exist as a 1 and a 0 at the same time, superposition allows for fractional values where the qubit exists as partially a 1 and partially a 0 in a probabilistic sense. Because of this fractional state, it's possible to design gates that work on the probability function rather than acting on single states of a bit. This flexibility allows for circuit designs to enhance algorithms, and such an approach to computing has shown gains in image classification. *Figure 14.1* shows the differences between a qubit and a bit:

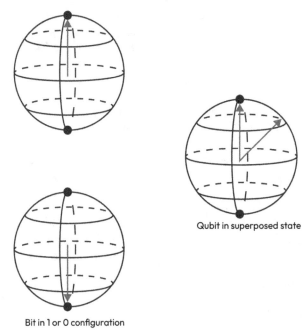

Qubit in superposed state

Bit in 1 or 0 configuration

Figure 14.1 – A visualization of qubits versus bits

Many types of machine learning algorithms have been translated for use on quantum computers to speed up solution convergence or scale algorithms to larger problems. Quantum machine learning is an active area of research, and network science algorithms are one of the larger areas of research within quantum machine learning. Let's explore two quantum network science algorithms in more detail.

Graph coloring algorithms

Graph coloring algorithms are a branch of network science algorithms aimed at finding the chromatic number of a graph. A **chromatic number** is the minimum number of colors required to color the vertices of a network such that no colors are adjacent to the same color. In real-world applications, chromatic numbers are important in determining the number of channels to allocate in communication systems or scheduling jobs within parallel processing. The minimum number of colors required to color the vertices of a network represents the minimum number of channels needed or the minimum number of jobs needed to complete the aforementioned tasks. *Figure 14.2* shows a small network with the colors shown:

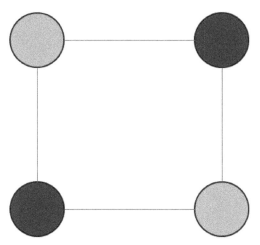

Figure 14.2 – A small network with four vertices, four edges, and a chromatic number of two

While *Figure 14.2* has an easy-to-determine chromatic number, computing the chromatic number of a network, in general, is thought to be **NP-hard**, meaning that it is not solvable in polynomial time. Algorithms that have high complexity translates to computation that is very difficult or currently impossible for large enough or dense enough networks. This poses problems in terms of solving problems in communication and job scheduling where the volume of calls or jobs is very large.

Fortunately, quantum graph coloring algorithms allow algorithms to run much faster than they do on classical machines, leading to quicker solutions for large graphs and potentially computable solutions for very large graphs. However, current quantum hardware systems limit the size of a graph on which a quantum graph coloring algorithm can be run, as the problem must be mapped to the physical system. As quantum hardware supports larger and larger networks, chromatic number computation will be feasible for more real-world problems.

Max flow/min cut

In *Chapter 4*, we explored **transportation logistics** and the **max flow/min cut algorithm**. Recall that this algorithm is useful in partitioning flows on networks and that it can be used to plan out road closures on a traffic grid, such as the grid shown in *Figure 14.3*:

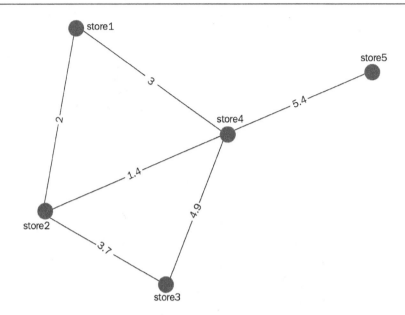

Figure 14.3 – A traffic grid in relation to five stores (considered in Chapter 4)

Finding the optimal cut to avoid service delays to each of these five stores (shown in *Figure 14.3*) is a difficult problem to compute, and computationally efficient solutions provide a good way to scale these types of analyses.

Another common use of the max flow/min cut algorithm is image partitioning, whereby the foreground and background of an image are determined. Image enhancement typically relies on distinguishing different parts of focus within the image. Photoshopping images also requires a foreground/background discernment step.

As with chromatic number calculations, computing max flow/min cut optima on large or dense graphs can be challenging when dealing with classical systems. Quantum versions of this algorithm exist and can optimize solutions more easily than classical versions. However, hardware size currently limits the application of quantum max flow/min cut algorithms on large problems.

Many other quantum versions of network algorithms exist and have given strong results on small problems, representing a solid future direction for scaling network algorithms in terms of newer approaches to hardware. As hardware development progresses, it's likely that network science will benefit, and the algorithms in network science that currently don't scale as well as other network science algorithms will find a path forward for difficult problems. So far, network science algorithms represent a major part of quantum machine learning research, and given their success, it's likely that network science and quantum computing will continue to influence each other.

Let's turn our attention to deep learning architectures, which can be formulated in both classical computing and quantum computing.

Neural network architectures as graphs

In *Chapter 1*, we touched on deep learning models, particularly in the context of generative artificial intelligence. Deep learning models surface in many areas of analytics, including **natural language processing**, **image classification**, and **time series forecasting**. Let's explore these applications in more detail.

Natural language processing is ubiquitous in the era of big data. Surveys often employ free text collection methods. Customers provide text reviews of products. Social scientists jot down notes when they are observing populations qualitatively (dubbed **ethnographic research**). Bloggers post regular content to disseminate ideas. Legal, medical, and educational notes can include biased content that needs to be addressed to protect people from institutional bias that limits future opportunities and wellness.

All this data needs to be parsed before it is fed into a classification model or exploratory data tools. Most of the tools that do this rely on deep learning models that can connect pieces of information separated across a sentence or a paragraph, which are termed **recurrent neural networks**. Many of the common packages in Python rely on pretrained recurrent neural networks, where the training steps have already been performed on a large dataset and saved as a model. Updating the pretrained models often involves adjusting the neural network weights and/or connections based on new data, perhaps from a specific domain of interest such as chemistry or law. *Figure 14.4* shows a simple recurrent neural network architecture. In practice, recurrent neural networks can have many more hidden layers and nodes with connections between them:

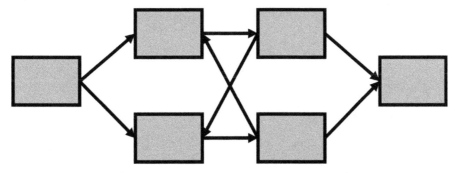

Figure 14.4 – A simple recurrent neural network diagram, where the hidden layers
provide feedback to each other in a forward and backward manner

Another key task in analytics these days is image classification. Social media allows users to upload high volumes of images across a dizzying array of topics that need to be tagged with topic information and flagged if it is not safe content. Trap cameras and drones allow for the surveillance of wildlife and potential poachers in protected areas, and these images need to be analyzed to protect wildlife and divvy up limited resources. Farmers can now monitor crops using robotic imaging technologies, which then match the images to disease classification models for the early identification of disease or distress in crops that endanger local food sources or hurt local economies. *Figure 14.5* shows three images that may be used in a training set of an image classifier to identify different types of animals at Kruger National Park in South Africa:

Figure 14.5 – Three images taken at Kruger National Park that could
be used within a training sample for an image classifier

From *Figure 14.5*, note that the images obtained from trap cameras or surveys may not include images of the entire animal, capture multiple animals within an image, have branches or other clutter obscuring the image, or truncate the full animal in the shot. These all pose issues to image classification and can increase sample size needs or classifier architecture complexity. Let's examine some potential solutions to these issues within deep learning.

Many image classification models utilize a deep learning architecture called **convolutional neural networks (CNNs)**. CNNs include many processing layers and a few pooling layers, where the features derived in prior layers are summarized and fed into subsequent layers. This hones in on the most relevant features found in the training process. Just as in natural language processing, many pretrained models exist that can be updated with new data to modify the neural network weights and connections according to the new data.

Now that we know a bit about deep learning models, let's dive into some of the architectures that are common in deep learning models and how these can be represented as networks.

Deep learning layers and connections

Despite the existence of many deep learning algorithms, they all include multiple processing layers consisting of vertices that are connected by edges. As the algorithm fits layers and connections to optimize its performance on training and test data, the network structure of the deep learning architecture evolves. This means that network science and its related tools, such as those found in the study of simplicial complexes, can be used to guide the fitting process or evaluate the algorithm at certain benchmarks. While this is a relatively new approach to deep learning algorithms, the approach has improved model fit, reduced the sample size needed to get a good fit, and sped up training.

Within image generation, the classification of filtered simplicial complex features—using an algorithm called **persistent homology**—has played an important role in the realistic generation of landscape terrain features and medical features (such as blood vessel branching) through deep learning algorithms called **generative adversarial networks**, which we briefly overviewed in *Chapter 1* as the basis for image generators. The combination of network-based metrics in the loss function upon which the network fitting is measured coaxes the algorithm into more realistic image generation.

Artificial image sets are very important within the medical context. Building image classifiers relies on large sample sizes, which might not exist for rare diseases or within small medical systems. Data augmentation through image generation allows us to create larger sample sizes with (hopefully!) realistic images upon which to build other applications, such as image classifiers. This, in turn, allows researchers to glean insight into the etiology of diseases by, say, studying morphological changes in cellular structure in different stages of a disease or identifying aggressive diseases as quickly as possible by matching patient samples to those curated by an algorithm trained on augmented prior patient data.

Analyzing architectures

Another perspective on deep learning architectures as network structures involves exploring good architectures by analyzing them using network science tools. **Large language models** (**LLMs**) are massive deep learning algorithms that currently involve billions to trillions of parameters to estimate across layers and their connections. They are trained on massive volumes of text data and require high volumes of computing resources to fit the model. Larger models tend to perform better than smaller models, necessitating more data and more computational power. This is not ideal, as computing resources have limitations based on hardware and substantial environmental impacts.

Network science provides a potential avenue for finding good, small LLM architectures that require less training data and fewer computational resources to fit. These architectures are more accessible to developers in resource-scarce countries when building models for underrepresented languages, and they offer a path toward sustainable LLM growth.

Coupling this network science approach with training data can reduce the training sample size needed to build an effective LLM, as well. Related documents with similar linguistic structure or content can be sampled from each structure or domain needed within the LLM. This provides comprehensive data that can be quality-checked using fewer resources. Better training data enhances model fitting, reduces training time, and reduces the computational resources needed to fit a good model. Now that we understand a bit about neural network architecture, let's turn to a type of network that integrates information about hierarchy and layering within its structure.

Hierarchical networks

In *Chapter 11*, we encountered ontologies and phylogenies within the linguistic and genetic context. A **hierarchical network** builds on the intuition within ontologies and phylogenies; hierarchical networks are tree-like networks that can include multiple sources and sinks flowing from the top of the tree to the bottom of the tree. Gene regulatory networks often follow this hierarchical network structure rather than strictly a tree structure, where multiple sources and sinks do not exist in the network, or an ontology structure, where parent vertices don't typically exist. *Figure 14.6* shows an example of a hierarchical network:

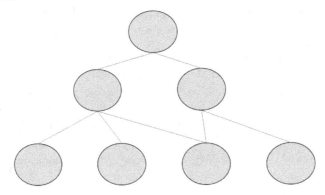

Figure 14.6 – A hierarchical network example where the second layer of
the network contains the parents of a vertex in the third layer

Studying gene regulation provides insight into myriad areas of research, including the evolutionary pathways within organ systems or species, the pathologies associated with gene dysregulation, and the discernment of cellular functions. Gene regulation isn't as simple as understanding the gene pathways themselves. Environmental conditions can change the structure of proteins associated with DNA or RNA to make DNA/RNA less or more accessible to the proteins involved in gene transcription. This, in turn, usually modifies the structures of those proteins further, either upregulating or downregulating the rate of transcription through accessibility to DNA/RNA.

Recent advances in the fields of **genomics** and **epigenomics** (which combines genetic analysis with environmental conditions) include measurement tools for gene expression and the structure of proteins around genes (which reflect local environmental conditions). By measuring these levels of expression under different conditions and different genetic mutation lines, it's possible to discern regulatory pathways within organisms.

Let's explore higher-order structures in network data from a more theoretical perspective before considering an example of hierarchical networks in snake venom epigenomics.

Higher-order structures and network data

If we think back to *Chapter 11*, language families start with a source—an initial language that evolves over time to other languages. This gives language families a hierarchical structure defined by time (and, typically, geography). Different levels emerge within different time periods, from the time of the original language to the modern-day derivatives of that language.

Within gene regulatory pathways, it's possible for multiple genes to influence those involved in later steps within the regulatory pathway, thus creating a more complicated network structure than a tree. However, order is still important, giving rise to a hierarchical network. Within biological studies, this structure can be hard to discern by using lab experiments, as the structure is not one-to-one, and, occasionally, the effects are not linear. However, recent technological advances allow researchers to discern these pathways more readily by varying the environment and/or knocking out the genes within a pathway.

Now that we know a bit about hierarchical networks, let's examine a real-world use case where these have shed light on evolutionary pathways and the variance in biological properties across different snake venoms. Venomous species are important for the health of many environments. In recent years, the study of their venom has led to pharmaceutical breakthroughs for human diseases, including non-addictive pain medications, treatments for cardiac diseases, and new antibiotics that can treat drug-resistant strains. Understanding differences across venomous species allows researchers and clinicians to develop better treatments for snake bite victims, allows medical researchers to study venom properties that may be useful in the pharmaceutical treatment of human disorders, and allows conservations to preserve biodiversity.

An example using gene families

Venomous species of snakes deplete venom either in a controlled way (striking prey) or an uncontrolled way (defensive bites when the snake is threatened). Eventually, the venom stores are depleted, and more venom must be produced for the snake to survive. Venom is mostly composed of proteins and polypeptides; given this, rapid venom production induces stress on the **endoplasmic reticulum** (**ER**), an organelle mainly involved in protein production within a cell. This suggests that the gene families involved in ER regulation may play a role in venom regulation and production.

A recent functional genomics study of the prairie rattlesnake revealed two distinct pathways heavily involved in venom regulation:

- An extracellular pathway related to the kinase pathway
- An ER-related protein pathway

This is logical, given that the signals would need to respond to exterior conditions (lower concentrations of venom) and interior conditions (overworked ERs); a balance between venom concentration and cellular health makes sense in terms of start and stop signals to regulate how much venom is being produced. Interestingly, these pathways are ubiquitous across species, and it is thought that similar pathways may be involved in general venom regulation.

Let's examine the phylogenic tree for genetic conservation within one pathway. You can find the image in the Perry 2022 paper cited in the reference section (*Figure 6C*). We'll discuss this image and its implications for evolutionary pathways in the following sections.

This tree suggests that the enhancer region studied in this pathway is conserved for venomous snake species, with a few further mutations occurring along the evolutionary pathway, giving rise to different types of venom with different modes of action within the body of an envenomated prey or predator.

Hierarchical networks allow us to organize information across levels, and visualizations such as *Figure 6C* from the Perry 2022 paper allow us to examine other relevant information (such as time) related to the organizational principles in a hierarchical network. However, hierarchical networks may not be flexible enough for some visualizations, and in the next section, we'll look at an even more flexible visualization tool.

Hypergraphs

Another exciting avenue of research in network science involves **hypergraphs**, which are an extension of networks in which multi-way relationships can include multiple vertices. In this way, the concept of edges is extended in a similar way to how the concept of edges was extended in *Chapter 6*, within the *Extending network metrics for time series analytics* section on simplicial complexes. In fact, the abstraction of simplicial complexes is one type of hypergraph.

Hypergraphs are often useful for visualizing database diagrams and planning information retrieval systems, where many-to-many relationships often exist. Some distributed computing systems even provide hypergraph algorithms to aid in these retrieval tasks at scale.

Visualization via hypergraphs is also used extensively in bioinformatics, where multi-way relationships exist between proteins, biological pathways, gene expression datasets, and metabolic processes. In complex systems, it is often easier to show relationships visually than to show a matrix or a list of the relationships. It's also much more compact to use this representation as opposed to diagrams with many overlapping lines or complicated tables.

Let's dive into the particulars of hypergraph visualization and then move on to a real-world example of hypergraphs simplifying data visualization within bioinformatics.

Displaying information

In science, it's important to communicate clearly and concisely, particularly when discussing complicated or nuanced concepts that may be very unfamiliar to readers. Accurate and concise infographics provide readers with alternative representations of material compared to long, jargon-laden paragraphs. This makes the material more accessible to a wider audience. For a new algorithm in machine learning or a new measurement tool in bioinformatics, the difference between a cool tool that is rarely adopted and a tool that people leverage to solve real problems is often a matter of marketing. Solid communication that invites readers into the material results in a larger and more diverse audience.

Let's consider four different ways of communicating a biological network's complex relationships to an audience. The first way is simply through text. Perhaps there are five pathways that overlap at certain stages within the pathway. **Pathway 1** follows its own pathway until merging with **Pathways 2** and **3,** which initially overlapped but then diverged after merging with **Pathway 1**. **Pathway 4** emerges on its own and then connects with **Pathway 2**. **Pathway 5** emerges on its own and then connects to **Pathway 3**. Can you picture this in your head as you read the text? It's difficult to visualize all of these pathways through text.

We could also show this via a table of relationships. For five pathways, this isn't terrible to visualize when using a pathway table, such as *Table 14.1*. However, consider what a table like this would look like for a complicated metabolic process or gene network that includes dozens or hundreds of pathways. Visualizing those through a table is not ideal:

	Pathway 1	Pathway 2	Pathway 3	Pathway 4	Pathway 5
Pathway 1		X	X	X	X
Pathway 2	X			X	
Pathway 3	X				X
Pathway 4	X	X			
Pathway 5	X		X		

Table 14.1 – A table showing pathway overlap across five metabolic pathways within an organism

Let's now consider a network with each pathway represented as a vertex and each connection through the metabolic pathways as an edge. While the limited number of pathways doesn't create a terribly messy graph, this representation will not scale, particularly as the pathway edges scale with the addition of more vertices. When reaching a dozen pathways, it would be very difficult to discern which edges connect which vertices by using this approach. *Figure 14.7* shows the metabolic pathway example visualized as a network:

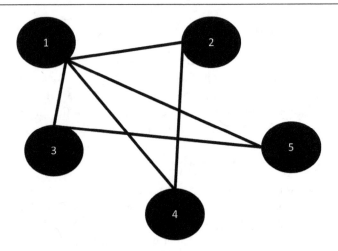

Figure 14.7 – A network representation of the relationships in metabolic pathways

Finally, let's consider a hypergraph representation of these pathways and their overlap. This approach looks a bit like a Venn diagram showing set overlaps, with different colors or textures parsing out the relationships into sets of related items. If given a temporal component to the data, we could even create evolving hypergraph representations for each branch point of the metabolic pathways. This is much easier to convey to a reader as the number of pathways increases, and considering this evolving network as a set of items allows for easier computation in terms of the changes and properties across evolving sets. *Figure 14.8* shows a hypergraph representation with temporal components included:

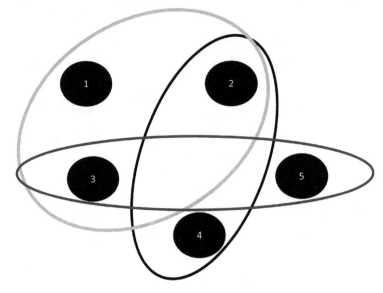

Figure 14.8 – A hypergraph representation of the metabolic pathways across time

Figure 14.8 captures both pathway-merging information and a temporal aspect as the pathways run to completion. This is much easier to understand than the text or the table representations and will scale more easily than a simple network diagram. Hypergraphs are useful visualization tools that scale well for complex problems.

Metadata

Oftentimes, networks include metadata about edges or vertices in the network, which are typically stored in tables. As we mentioned earlier in this section, hypergraphs are useful tools when wrangling complex databases. In fact, hypergraph databases exist and can be quite helpful in the data management of OWL databases, language topic models, document storage, and much more. The flexibility of hypergraph databases compared to graph databases includes the ability to easily capture and represent multi-way connections and interactions between data tables, documents, or sets of tables or documents.

In returning to our bioinformatics example, hypergraph databases are ideal for managing bioinformatics data related to genes, proteins, metabolic pathways, or other interconnected sets of processes. The layers can represent the different aspects of pathway overlap, different periods of time, or different organisms with overlapping metabolic or regulatory pathways.

Consider the venom example in the previous section. A hypergraph database could contain all the known gene pathways associated with venom production across venomous organisms. Phylogenic relationships, distribution geographies, antivenin classes, and other types of metadata can be used to connect groups of gene pathways to facilitate the study of particular genes or regulatory functions across habitats, species, and interactions with human populations. Updating this type of knowledge periodically as new papers are published would accelerate research in the field and tie in different branches of knowledge in a manner that is easy for researchers to access.

Given the utility of hypergraph databases, we are likely to see more development in this field of network science in the coming years.

Summary

In this chapter, we've pushed the limits of network science by merging network algorithms with quantum computing frameworks, enhanced our understanding of scaling the deep learning algorithms that are ubiquitous in modern data science, and explored special types of networks that are critical in molecular biology and genetics. Network science plays an important role in data science, offering scalability and novel solutions to common problems. As the field of network science evolves, data science and its practitioners will continue to benefit. We hope this book will equip and inspire anyone who works with data to push the boundaries of knowledge and solve difficult problems in the world by using data. Come join us!

References

Berkolaiko, G., & Kuchment, P. (2013). *Introduction to quantum graphs* (No. 186). American Mathematical Soc.

Biamonte, J., Wittek, P., Pancotti, N., Rebentrost, P., Wiebe, N., & Lloyd, S. (2017). *Quantum machine learning. Nature, 549*(7671), 195-202.

Cerezo, M., Verdon, G., Huang, H. Y., Cincio, L., & Coles, P. J. (2022). *Challenges and opportunities in quantum machine learning. Nature Computational Science, 2*(9), 567-576.

Cui, S. X., Freedman, M. H., Sattath, O., Stong, R., & Minton, G. (2016). *Quantum max-flow/min-cut. Journal of Mathematical Physics, 57*(6).

Ekanayake, E. M. U. S. B., Daundasekara, W. B., & Perera, S. P. C. (2022). *New Approach to Obtain the Maximum Flow in a Network and Optimal Solution for the Transportation Problems. Modern Applied Science, 16*(1), 30.

Feng, S., Heath, E., Jefferson, B., Joslyn, C., Kvinge, H., Mitchell, H. D., ... & Purvine, E. (2021). *Hypergraph models of biological networks to identify genes critical to pathogenic viral response. BMC bioinformatics, 22*(1), 1-21.

Iordanov, B. (2010). *Hypergraphdb: a generalized graph database.* In *Web-Age Information Management: WAIM 2010 International Workshops: IWGD 2010, XMLDM 2010, WCMT 2010, Jiuzhaigou Valley, China, July 15-17, 2010 Revised Selected Papers 11* (pp. 25-36). Springer Berlin Heidelberg.

James, D. F., Kwiat, P. G., Munro, W. J., & White, A. G. (2001). *Measurement of qubits. Physical Review A, 64*(5), 052312.

Karlebach, G., & Shamir, R. (2008). *Modelling and analysis of gene regulatory networks. Nature reviews Molecular cell biology, 9*(10), 770-780.

Leighton, F. T. (1979). *A graph coloring algorithm for large scheduling problems. Journal of research of the national bureau of standards, 84*(6), 489.

Perry, B. W., Gopalan, S. S., Pasquesi, G. I., Schield, D. R., Westfall, A. K., Smith, C. F., ... & Castoe, T. A. (2022). *Snake venom gene expression is coordinated by novel regulatory architecture and the integration of multiple co-opted vertebrate pathways. Genome Research, 32*(6), 1058-1073.

Titiloye, O., & Crispin, A. (2011). *Quantum annealing of the graph coloring problem. Discrete Optimization, 8*(2), 376-384.

Vidya, V., Achar, R. R., Himathi, M. U., Akshita, N., Kameshwar, V. H., Byrappa, K., & Ramadas, D. (2021). *Venom peptides–A comprehensive translational perspective in pain management. Current Research in Toxicology, 2*, 329-340.

Wittek, P. (2014). *Quantum machine learning: what quantum computing means to data mining.* Academic Press.

Index

packtpub.com

Subscribe to our online digital library for full access to over 7,000 books and videos, as well as industry leading tools to help you plan your personal development and advance your career. For more information, please visit our website.

Why subscribe?

- Spend less time learning and more time coding with practical eBooks and Videos from over 4,000 industry professionals

- Improve your learning with Skill Plans built especially for you

- Get a free eBook or video every month

- Fully searchable for easy access to vital information

- Copy and paste, print, and bookmark content

Did you know that Packt offers eBook versions of every book published, with PDF and ePub files available? You can upgrade to the eBook version at packtpub.com and as a print book customer, you are entitled to a discount on the eBook copy. Get in touch with us at customercare@packtpub.com for more details.

At www.packtpub.com, you can also read a collection of free technical articles, sign up for a range of free newsletters, and receive exclusive discounts and offers on Packt books and eBooks.

Other Books You May Enjoy

If you enjoyed this book, you may be interested in these other books by Packt:

Hands-On Graph Neural Networks Using Python

Maxime Labonne

ISBN: 978-1-80461-752-6

- Understand the fundamental concepts of graph neural networks
- Implement graph neural networks using Python and PyTorch Geometric
- Classify nodes, graphs, and edges using millions of samples
- Predict and generate realistic graph topologies
- Combine heterogeneous sources to improve performance
- Forecast future events using topological information
- Apply graph neural networks to solve real-world problems

Graph Machine Learning

Claudio Stamile, Aldo Marzullo, Enrico Deusebio

ISBN: 978-1-80020-449-2

- Write Python scripts to extract features from graphs
- Distinguish between the main graph representation learning techniques
- Learn how to extract data from social networks, financial transaction systems, for text analysis, and more
- Implement the main unsupervised and supervised graph embedding techniques
- Get to grips with shallow embedding methods, graph neural networks, graph regularization methods, and more
- Deploy and scale out your application seamlessly

Packt is searching for authors like you

If you're interested in becoming an author for Packt, please visit `authors.packtpub.com` and apply today. We have worked with thousands of developers and tech professionals, just like you, to help them share their insight with the global tech community. You can make a general application, apply for a specific hot topic that we are recruiting an author for, or submit your own idea.

Share Your Thoughts

Now you've finished *Modern Graph Theory Algorithms with Python*, we'd love to hear your thoughts! Scan the QR code below to go straight to the Amazon review page for this book and share your feedback or leave a review on the site that you purchased it from.

`https://packt.link/r/1-805-12789-6`

Your review is important to us and the tech community and will help us make sure we're delivering excellent quality content.

Download a free PDF copy of this book

Thanks for purchasing this book!

Do you like to read on the go but are unable to carry your print books everywhere?

Is your eBook purchase not compatible with the device of your choice?

Don't worry, now with every Packt book you get a DRM-free PDF version of that book at no cost.

Read anywhere, any place, on any device. Search, copy, and paste code from your favorite technical books directly into your application.

The perks don't stop there, you can get exclusive access to discounts, newsletters, and great free content in your inbox daily

Follow these simple steps to get the benefits:

1. Scan the QR code or visit the link below

https://packt.link/free-ebook/9781805127895

2. Submit your proof of purchase
3. That's it! We'll send your free PDF and other benefits to your email directly

www.ingramcontent.com/pod-product-compliance
Lightning Source LLC
LaVergne TN
LVHW081519050326
832903LV00025B/1545